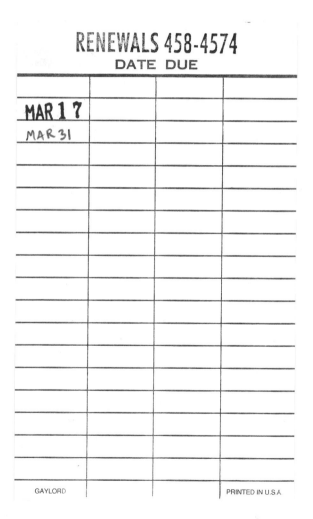

RENEWALS 458-4574

DATE DUE

MAR 17			
MAR 31			
GAYLORD			PRINTED IN U.S.A.

NUMERICAL THERMAL ANALYSIS

by

Satish P. Ketkar, Ph.D.

The MacNeal-Schwendler Corporation

ASME Press • • New York • • 1999

Copyright © 1999 by The American Society of Mechanical Engineers
Three Park Avenue, New York, NY 10016

Library of Congress Cataloging-in-Publication Data

Ketkar, Satish P., 1960—
 Numerical thermal analysis/by Satish P. Ketkar
  p.   cm.
 Includes bibliographical references and index.
 ISBN 0-7918-0073-3
 1. Heat engineering—Mathematics.   2. Numerical analysis.
 I. Title.
 TJ260-.K38   1999
  98-32285
 621.402—dc21
  CIP

This book is dedicated to my children,
Saarika and Sachin.

Table of Contents

Preface

This book is intended to provide a bridge between the academic world and the industrial world in the area of computer-aided numerical thermal analysis. It is written more as a primer than as an academic textbook. Basic concepts of computational thermal engineering are addressed. Chapters on the finite difference and control volume methods are compiled from various thermal science courses I took while working on a doctoral degree at the University of Tennessee, Knoxville. Finite element method covered in Chapter 5 is based partly on my study notes compiled during a home correspondence course called "Finite Element Heat Transfer" offered by the American Institute of Aeronautics and Astronautics (AIAA). The final chapter discusses a unique hybrid method that combines the best features of finite element modeling and the computational efficiency of finite difference network solution techniques. I came across this method while working at PDA engineering and later at MacNeal-Schwendler Corporation in software development of P/Thermal, training, and technical applications support capacities. This unique and robust technique is used in commercially available code MSC/Patran/Thermal™ software.

There are many books written on numerical heat transfer. These books address finite difference, finite element, and control volume methods. However, few books cover the basics of all these numerical methods. The topics covered in the traditional textbooks are in many instances too academic and/or too mathematical. Furthermore, I have not come across any book that addresses the hybrid method. A discussion on the hybrid method is scattered in the literature, with various portions appearing in technical papers and product manuals. The desire to present this unique method in a coherent form was one of the strong motivations for my undertaking the writing of this book. The books currently on the market treat the subject of conduction-based thermal analysis as a subset of the generalized topic of numerical heat transfer and expend a lot more ink on computational fluid dynamics (CFD). In industry, CFD is still considered a specialized area, and many more thermal engineers get their jobs done by using computerized tools for thermal analysis. Since the majority of these tools are conduction-based, convection and radiation is introduced only as a boundary condition. During my years in industry, I taught thermal analysis courses for a variety of engineers in automotive, defense, aerospace, power generation, and consumer products industries. I have found that there was no single book on the market that I could recommend as a quick desktop reference for engineers who use computerized thermal analysis tools. This book is designed to fill this need. It can be used by students of numerical heat transfer who are about to embark on industrial careers, as well as by practicing engineers who want to refresh their fundamentals of numerical thermal analysis.

I would like to thank Richard Haddock of PDA Engineering and Mac-Neal Schwendler Corporation for many stimulating discussions over the years regarding thermal analysis. I also thank Mike Chainyk of MacNeal-Schwendler Corporation for reviewing the manuscript and making useful suggestions.

Thanks are also due to the reviewers of the ASME Press for their good suggestions. Any shortcomings that may remain are my own fault, and any suggestions for improvement are welcome.

Satish P. Ketkar, Ph.D.
The MacNeal-Schwendler Corporation
Southfield, Michigan
January, 1999

Chapter 1

Introduction

Thermal analysis has grown and matured to the point where there are a number of commercial and public-domain computer programs available to help engineers in the thermal design and analysis process. To achieve concurrent engineering, industries increasingly rely on computer-aided engineering (CAE) to shorten design cycle times, increase product quality and reliability, and reduce prototyping and testing. The engineer no longer has the luxury of overdesigning the product to avoid thermally induced failures. The demands of modern competitive design require sophisticated techniques. As the number of commercially available computer programs is also on the increase, the job of selecting the right one for the application is very critical. It is a dangerous practice to select and use computer technology without understanding the underlying principles on which it is based. Selecting an inappropriate program for the application can lead to thousands of wasted dollars not to mention frustration.

Many books have been written on numerical heat transfer analysis. They address finite difference, finite element, control volume, and to a lesser degree, boundary element approaches to solving the governing thermal equations. However, very few books cover the basics of all these numerical methods. Also, the discussions of these topics in traditional books are generally too academic or too mathematical. Designed as a primer, rather than as an academic textbook, this book is intended to be used by practicing engineers as a quick reference guide for the numerical methods used in thermal analysis, as well as by students about to enter the industrial world.

Finite difference method (FDM) is the most widely used method in thermal analysis. Its popularity is due to the fact that the mathematical concepts of approximating a continuous domain with a network of discrete points (called a mesh) is relatively simple. The partial derivatives of the governing heat conduction equation are replaced by finite differences using the Taylor series. In general, first and second derivatives are estimated using second-order difference approximations.

Control volume, or finite volume, technique is based on integral equation of the governing differential equation over the control volumes. The calculation domain is subdivided into a number of nonoverlapping control volumes (mesh). Both finite difference and finite volume methods lead to similar discretization equations, which can be solved by matrix methods on a digital computer. Both methods require that the computational domain be discretized by using orthogonal structured meshes. For problems with irregular (geometric) domains, one can use complex variable transformations to change the domain to an orthogonal one or use fine discretization and stair-stepping technique near the curvature.

1

Finite element method (FEM) was originally developed for solving structural problems. The method has been extended to address nearly all classes of problems, including thermal and fluid flow problems. The finite element method is based on integral minimization of error, which is quite unlike the finite difference method based on Taylor series expansion. The main advantage of the finite element method is that it allows for unstructured meshes and, hence, highly irregular geometry (domains) can be handled as there is no need for complex coordinate or variable transformations.

In the boundary element method (BEM), one transforms the problem using Green's theorem from a volume integral to a surface integral for three-dimensional problems or from a surface integral to a line integral for two-dimensional problems. As the dimensionality of the problem is reduced, one only needs to discretize the outer boundary of the domain. The method has been effective in linear problems governed by Laplace's or Poisson's equation. Because the method has limited scope, it will not be considered in this work.

A recent addition to the numerical thermal analysis methods is the novel connectivity and resistor equivalent (CARE) technique. CARE combines the ease of modeling by finite element method with the robustness and efficiency of thermal network solution methodologies. This technique is used in the commercially available state-of-the-art thermal analysis software MSC/Patran/Thermal™.

Any engineering analysis has three major steps: (1) input, (2) compute, (3) output. Input and output steps are popularly known in the engineering analysis industry as preprocessing and postprocessing. Every analysis requires the engineer to define the computational domain. This is achieved by first creating the geometry of the domain in a computer-aided design (CAD) system or computer-aided engineering (CAE) system. Once the geometry is defined, one paves it with a mesh. The next step is to define boundary conditions, physical and material properties, and analysis parameters. The model is then submitted for analysis, which is the compute stage. After successfully crunching the numbers, one graphically displays the results or the solution; this is called postprocessing. Pre- and postprocessing has evolved into a discipline of its own right. Sophisticated packages, such as MSC/Patran™ and SDRC/Ideas™, are commercially available. These packages let the engineer create models and display the results of the simulation in a variety of ways.

In summary, this book addresses the basic principles of numerical thermal analysis. Finite difference, finite volume, and finite element techniques are described and the pros and cons of each method are discussed. A unique resistor-equivalent (CARE) network technique is discussed. The CARE network is a mathematically exact resistor equivalent to the Galerkin finite element. This unique FE-to-FD hybrid approach makes applying complex thermal boundary conditions easy and at the same time gives the analyst the flexibility of creating models with the finite-element-based pre- and postprocessors.

Chapter 2

Heat Conduction: Fundamentals and Governing Equations

Whenever a temperature gradient exists in a system or when two bodies at different temperatures are brought into contact, energy transfer takes place. This energy transport process is known as heat transfer. Heat transfer can be classified into three distinct modes—conduction, convection, and radiation.

HEAT CONDUCTION

Conduction is a process by which heat flows from a region of high temperature to a region of low temperature within a medium (solid, liquid, or gas) or between mediums that are in direct physical contact. The energy is transmitted by direct molecular communication without appreciable displacement of molecules.

The basic law that relates the heat flow to the temperature gradient was proposed by the French physicist, Fourier. It states that for a homogeneous solid, the heat flux (heat per unit area) is proportional to the gradient of temperature, as follows

$$\mathbf{q} = Q/A \propto (\text{Grad } T) \tag{2.1}$$

where Q is heat flow in watts or Btu/hr
  $\mathbf{q}$ is heat flux (heat flow per unit area)
  A is cross-sectional area
  T is temperature

Grad is the gradient operater, which is defined as

$$\text{Grad} = \frac{\partial}{\partial x}\,\hat{i} + \frac{\partial}{\partial y}\,\hat{j} + \frac{\partial}{\partial z}\,\hat{k}$$

Using k as the constant of proportionality, one can rewrite Eq. (2.1) as

$$\mathbf{q} = Q/A = k\,\text{Grad } T$$

where k is called the isotropic thermal conductivity (thermal conductivity is independent of direction).

As the heat flows in the direction of decreasing temperature, the gradient of temperature (Grad T) is negative. As the heat flow becomes a positive value, the Fourier law is written as [1, 21, 22]

$$\mathbf{q} = -\mathbf{k}(\text{Grad } T)$$

where $\mathbf{q}$ is the heat flux vector, $\mathbf{k}$ is the thermal conductivity vector, and T (the temperature) is a scalar quantity. The three components of the heat flux vector in Cartesian x, y, z coordinate system can be written as

$$\mathbf{q}_x = -\mathbf{k}_x \frac{\partial T}{\partial x} \qquad \mathbf{q}_y = -\mathbf{k}_y \frac{\partial T}{\partial y} \qquad \mathbf{q}_z = -\mathbf{k}_z \frac{\partial T}{\partial z} \qquad (2.2)$$

Thermal conductivity of the material is an important property that controls the rate of heat flow in a medium. Thermal conductivities have a wide range of values for engineering materials. Thermal conductivity is high for pure metals and is quite low for gases and vapors. Listed below are some typical thermal conductivity values [1]:

Metals	50–415 W/m°C
Alloys	12–120 W/m°C
Nonmetallic liquids	0.17–0.7 W/m°C
Insulators	0.03–0.17 W/m°C
Gases	0.007–0.17 W/m°C

In practice, thermal analysis is inherently nonlinear, as thermal conductivity varies with temperature. It decreases with temperature for most pure materials, while it increases with temperature for gases and insulating materials. At low temperatures, thermal conductivity is a strong function of temperature, i.e., it varies rapidly with temperature. The reader is referred to references 1 and 2 for thermal conductivities of various materials.

HEAT CONDUCTION EQUATION

The main objective of thermal analysis is to determine the temperature distribution in the model (medium) based on the environmental conditions imposed on its boundaries. Once the temperature distribution is known, heat flow at any point in the model or on any of its boundaries can be computed from Fourier's law of heat conduction. Temperature distribution can be used to calculate thermal stresses, or it can be used to optimize heat loss or gain.

To derive the equation that governs the temperature distribution $T(x, y, z, t)$, consider an infinitesimal control volume $dx\,dy\,dz$ whose faces are parallel to the global Cartesian axes x, y, z.

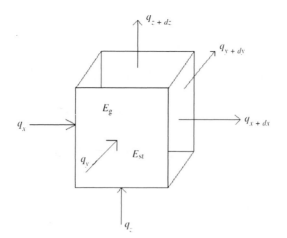

FIGURE 2.1. Heat balance on a control volume.

Assume the rate of heat generation in the volume is $\dot{q}$ (watts/m^3). The total heat generation (E_g) inside the control volume is

$$E_g = \dot{q}(dx\ dy\ dz)$$

Some energy will be stored inside the control volume

$$E_{st} = \text{density} \cdot \text{specific heat} \cdot dx\ dy\ dz = \rho \cdot C_p \cdot dx\ dy\ dz$$

Applying the first law of thermodynamics, the conservation of energy requires that (Rate of energy entering the control volume) − (The rate of energy leaving the control volume) + (Rate of energy generated in the control volume) = (Rate of energy stored in the control volume) or

$$q_x + q_y + q_z - q_x + dx - q_y + dy - q_z + dz + \dot{q} \cdot dx\ dy\ dz$$

$$= \rho C_p(\partial T/\partial t)dx\ dy\ dz \tag{2.3}$$

$$(-\partial q_x/\partial x)dx - (\partial q_y/\partial y)dy - (\partial p_z/\partial z)dz + \dot{q} \cdot dx\ dy\ dz$$

$$= \rho C_p(\partial T/\partial t)dx\ dy\ dz \tag{2.4}$$

Applying Fourier's Law of heat conduction results in

$$q_x = -k_x(\partial T/\partial x)dy\ dz$$
$$q_y = -k_y(\partial T/\partial y)dx\ dz$$
$$q_z = -k_z(\partial T/\partial z)dx\ dy$$

And dividing by the volume $dx\ dy\ dz$, we get

$$\frac{\partial}{\partial x}\left(k_x\,\frac{\partial T}{\partial x}\right) + \frac{\partial}{\partial y}\left(k_y\,\frac{\partial T}{\partial y}\right) + \frac{\partial}{\partial z}\left(k_z\,\frac{\partial T}{\partial z}\right) + \dot{q} = \rho C_p\,\frac{\partial T}{\partial t} \tag{2.5}$$

Equation (2.5) is the Cartesian form of the heat conduction equation. The solution of the heat conduction equation gives the temperature distribution in the medium.

For isotropic medium and constant thermal conductivity k, the equation is

$$\nabla^2 T + (\dot{q}/k) = 1/\alpha \cdot (\partial T/\partial t) \tag{2.6}$$

where

$$\nabla^2 \equiv (\partial^2/\partial x^2) + (\partial^2/\partial y^2) + (\partial^2/\partial z^2)$$

and

$$\alpha \equiv k/\rho Cp,\ \text{the thermal diffusivity}$$

If there is no heat source present, the heat equation reduces to

$$\nabla^2 T = 1/\alpha \cdot (\partial T/\partial t) \qquad \text{Fourier equation} \tag{2.7}$$

If the model is in steady state, with heat generation and constant properties, then the applicable form of the equation is

$$\nabla^2 T + (\dot{q}/k) = 0 \qquad \text{Poisson's equation} \tag{2.8}$$

For steady state, no heat generation, and temperature-independent properties, the equation is

$$\nabla^2 T = 0 \qquad \text{Laplace's equation} \tag{2.9}$$

BOUNDARY CONDITIONS

The heat conduction equation can have infinitely many solutions unless initial conditions (for transient problems) and boundary conditions are imposed. The initial condition imposes a temperature distribution in the medium at time = 0, and the boundary conditions specify the heat flow or the temperature at the boundaries of the region. Boundary conditions can be classified into three categories, as noted below.

Dirichlet Boundary Condition: The Dirichlet boundary condition is also known as the boundary condition of the first kind. A temperature is prescribed along a portion of the boundary. The temperature can be fixed for all times, or it can be a function of time $T(x, y, z, t)$.

Neumann Boundary Condition: The Neumann boundary condition is also known as the boundary condition of the second kind. A normal derivative of the temperature (heat flux) is prescribed at a portion of the boundary as follows:

$$(\partial T/\partial n)_{\text{surface}} = f(x, y, z, t) = q'' \tag{2.10}$$

A zero prescribed temperature derivative or flux is the adiabatic, or insulated, surface boundary condition.

Mixed Boundary Condition: Mixed boundary condition is also known as the boundary condition of the third kind. For the mixed boundary condition, a combination of heat flux and temperature is specified at a portion of the boundary; it corresponds to convection/radiation heating or cooling at the surface and can be written as

$$-k(\partial T/\partial n)|_{\text{surface}} = q'' = h(T_s - T_\infty) + \epsilon\sigma(T_s^4 - T_\infty^4)$$

where $\epsilon \equiv$ the emissivity of the surface
$\sigma \equiv$ Stefan Boltzman constant
$h \equiv$ heat transfer coefficient
$T_x \equiv$ surface temperature
$T_\infty \equiv$ ambient temperature

Linearizing the radiation term, we get

$$-k(\partial T/\partial n)|_{\text{surface}} = h_c(T_s - T_\infty) + \epsilon\sigma(T_s + T_\infty)(T_s^2 + T_\infty^2)(T_s - T_\infty)$$
$$= h_c(T_s - T_\infty) + h_r(T_s - T_\infty) \tag{2.11}$$

where $h_c \equiv$ convective heat transfer coefficient
$h_r \equiv$ radiative heat transfer coefficient

Thermal analysis for real-world problems is inherently nonlinear. This is because thermal conductivity is a function of temperature, and the nonlinearity enters the fray through the governing partial differential equation. The nonlinearity can also play a part through boundary conditions. As seen earlier, the radiation boundary condition is highly nonlinear (the radiative heat transfer coefficient is a nonlinear function of temperature, and emissivity is a function of temperature). The convective heat transfer coefficient can also be a function of temperature. (Natural convection heat transfer is proportional to 5/4th power of the temperature difference.) If there are phase changes involved, then a seemingly linear problem becomes nonlinear since the location of the phase front is not known up front.

Analytical solutions available for nonlinear problems are very few, and those available are for very simple geometric configurations and/or simple boundary conditions. Therefore, one usually has to resort to numerical solutions for the problems encountered in the real world. In addition to describing the widely used methods for numerical thermal analysis mentioned earlier, a chapter on how to go about selecting and implementing a commercial thermal analysis software program is included.

There are many books written by heat transfer researchers that do great justice to the theoretical aspects of numerical thermal analysis. However, these books are too detailed, and practicing engineers may not have time to read them and understand all their content. In this work just enough of the theoretical basis is covered to permit the readers to understand the basic concepts, and the main emphasis is to enable readers to be thermal analysts.

Chapter 3

Finite Difference Method

The basic concept in the finite difference approach to the numerical solution of thermal problems is to replace the partial differential equation of heat conduction and its boundary conditions with finite difference approximations using the Taylor series expansion of the derivative about the point of interest. The solutions that are obtained are approximate in the sense that there is some difference between the numerical solution and the "exact" solution, if known. The motivation behind using numerical methods for partial differential equation of heat conduction is to be able to solve otherwise unsolvable problems. This means that one usually cannot compare the numerical solution with a known analytical solution. One can, however, test the numerical solution algorithm on problems with known analytical solutions.

Let $f(x)$ be a function that can be expanded in a Taylor series. Therefore,

$$f(x + \Delta x) = f(x) + f'(x)\Delta x + f''(x)(\Delta x^2/2) + f'''(x)(\Delta x^3/6) + O(\Delta x^4) \quad (3.1)$$
$$f(x - \Delta x) = f(x) - f'(x)\Delta x + f''(x)(\Delta x^2/2) - f'''(x)(\Delta x^3/6) + O(\Delta x^4) \quad (3.2)$$

Note that $'$ denotes the first derivative, $''$ denotes the second derivative, and so on. $O(\Delta x^4)$ means a term proportional to Δx^4. Adding Eqs. (3.1) and (3.2) gives

$$f(x + \Delta x) + f(x - \Delta x) = 2f(x) + f''(x)\Delta x^2 + O(\Delta x^4)$$

Therefore,

$$f''(x) = [f(x + \Delta x) - 2f(x) + f(x - \Delta x)]/\Delta x^2 + O(\Delta x^2) \quad (3.3)$$

Equation (3.3) is the finite difference (FD) approximation of the second derivative of function $f(x)$, and the truncation error is proportional to Δx^2. The expression is therefore said to be second-order accurate.

Subtracting Eq. (3.2) from Eq. (3.1), we get

$$f(x + \Delta x) - f(x - \Delta x) = 2f'(x)\Delta x + f'''(x)(\Delta x^3/3) + O(\Delta x^4)$$

Therefore,

$$f'(x) = \{f(x + \Delta x) - f(x - \Delta x)\}/2\Delta x + O(\Delta x^2) \quad (3.4)$$

Equation (3.4) is the central difference expression for the first derivative of $f(x)$. One can derive the expression for $f'(x)$ from just the use of Eq. (3.2) or (3.3):

$$f'(x) = [f(x) - f(x - \Delta x)]/\Delta x + O(\Delta x) \tag{3.5}$$

$$f'(x) = [f(x + \Delta x) - f(x)]/\Delta x + O(\Delta x) \tag{3.6}$$

Equation (3.5) is called the backward difference expression and Eq. (3.6) is called the forward difference expression for $f'(x)$. Equations (3.5) and (3.6) are only first-order accurate and are therefore discarded in favor of Eq. (3.4) when the space derivatives are approximated. Forward difference is usually used for discretizing the time derivative in the heat conduction equation. Second-order accurate expressions can be derived easily for forward and backward differences, and doing so is left as an exercise for the readers.

A similar approach is used for partial derivatives. To write the expression for $\partial f/\partial x$ or f_x in finite difference form, where $f \equiv f(x,y)$, y is held constant and $f(x,y)$ is expanded in a Taylor series in the x direction:

$$f(x + \Delta x, y) = f(x,y) + \frac{\partial f}{\partial x}(x,y)\Delta x + \frac{\partial^2 f}{\partial (x)^2}(x,y)\frac{\Delta x^2}{2}$$
$$+ \frac{\partial^3 f}{\partial (x)^3}(x,y)\frac{\Delta x^3}{6} + O(\Delta x^4) \tag{3.7}$$

$$f(x - \Delta x, y) = f(x,y) - \frac{\partial f}{\partial x}(x,y)\Delta x + \frac{\partial^2 f}{\partial (x)^2}(x,y)\frac{\Delta x^2}{2}$$
$$- \frac{\partial^3 f}{\partial (x)^3}(x,y)\frac{\Delta x^3}{6} + O(\Delta x^4) \tag{3.8}$$

Adding Eqs. (3.7) and (3.8), we get

$$\frac{\partial^2 f}{\partial (x)^2}(x,y) = \frac{f(x + \Delta x, y) - 2f(x,y) + f(x - \Delta x, y)}{(\Delta x)^2} + O(\Delta x)^2 \tag{3.9}$$

Subtracting Eq. (3.8) from Eq. (3.7) gives

$$\frac{\partial f}{\partial x}(x,y) = \frac{f(x + \Delta x, y) - f(x - \Delta x, y)}{2\Delta x} + O(\Delta x^2) \tag{3.10}$$

Expanding in the y direction, we get

$$\frac{\partial f}{\partial y}(x,y) = \frac{f(x, y + \Delta y) - f(x, y - \Delta y)}{2\Delta y} + O(\Delta y^2) \tag{3.11}$$

and

$$\frac{\partial^2 f}{\partial (y)^2}(x,y) = \frac{f(x, y + \Delta y) - 2f(x,y) + f(x, y - \Delta y)}{(\Delta y)^2} + O(\Delta y^2) \tag{3.12}$$

The extension to the third dimension is straightforward. Considering $f(x,t)$, the finite difference expression for function f with respect to time, we get

$$\frac{\partial f}{\partial t}(x,t) = \frac{f(x,t+\Delta t) - f(x,t-\Delta t)}{2\Delta t} + O(\Delta t^2) \qquad (3.13)$$

Usually, it is intuitive to march ahead in time from current time. Therefore, a forward difference approximation is given by

$$\frac{\partial f}{\partial t}(x,t) = \frac{f(x,t+\Delta t) - f(x,t)}{\Delta t} + O(\Delta t) \qquad (3.14)$$

which is less accurate than the central difference approximation given by Eq. (3.13).

STEADY STATE LINEAR HEAT CONDUCTION

To illustrate the finite difference concept, let us study a two-dimensional steady state heat conduction with constant properties and Dirichlet boundary conditions. The governing equation and the boundary conditions are as follows:

$$\frac{\partial^2 T}{\partial (x)^2} + \frac{\partial^2 T}{\partial (y)^2} = 0 \qquad (3.15)$$

with

$$T = 50°C \ @ \ x = 0$$
$$T = 200°C \ @ \ x = 1$$
$$T = 300°C \ @ \ y = 0$$
$$T = 100°C \ @ \ y = 1$$

Numerical solution is a discrete solution and is obtained at predetermined locations, or discrete points, in the region of interest. One has to partition the region of interest into a finite number of points for which the solution will be obtained. Let us discretize the square region into 3 equal segments in the x direction and 3 equal segments in the y direction (Fig. 3.1). The intersections of these fictitious lines are the points where the solution is obtained. These points are called "nodes." The temperatures on the boundary in this example are known; therefore, a solution is sought only at nodes 1, 2, 3, 4.

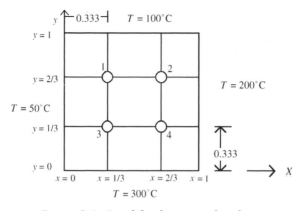

FIGURE 3.1. Partitioning a region into discrete points for numerical solution.

Replacing partial derivatives in Eq. (3.15) by their finite difference approximations, we get

$$\frac{T(x + \Delta x, y) - 2T(x, y) + T(x - \Delta x, y)}{(\Delta x)^2} + \frac{T(x, y + \Delta y) - 2T(x, y) + T(x, y - \Delta y)}{(\Delta y)^2}$$

$$+ O(\Delta x^2) + O(\Delta y^2) = 0 \qquad (3.16)$$

in the example $\Delta x = \Delta y = 0.3333$. Since $\Delta x = \Delta y$, we can simplify to

$$4T(x, y) = T(x - \Delta x, y) + T(x + \Delta x, y) + T(x, y + \Delta y) + T(x, y - \Delta y)$$

By converting into an index notation (i, j) instead of (x, y), where $x = (i - 1)\Delta x$ and $y = (j - 1)\Delta y$, we get

$$4T(i, j) = T(i - 1, j) + T(i + 1, j) + T(i, j - 1) + T(i, j + 1) \qquad (3.17)$$

Equation (3.17) is a finite difference representation of the differential equation governing heat conduction in the slab problem under study at point (i, j). If one writes Eq. (3.17) for every point (i, j) where a solution is sought, it leads to a system of algebraic equations for unknown values of T. Applying the boundary conditions eliminates some of the equations. The application of finite difference method to the heat equation leads to equations similar to Eq. (3.17). The equation states the relationship between adjacent nodes. The temperature at node (i, j) is influenced by temperatures at nodes $(i + 1)$, $(i - 1)$, $(j + 1)$, and $(j - 1)$. The strength of the influence is reflected in the value of the coefficients. To have a physically realistic solution (i.e., a solution that would not violate the second law of thermodynamics), all the coefficients on the right-hand side of Eq. (3.17) have to be the same sign as the coefficient of $T(i, j)$. The resulting set of algebraic equations is solved for unknown temperatures.

A variety of methods are available to solve a system of linear algebraic equations and are classified based on whether they are direct or iterative. Direct methods involve a fixed number of arithmetic operations and are suitable when the number of node points (unknown temperatures) is small. These methods involve excessive computer operations. For problems with large number of node points or with nonlinearities, iterative methods of solution may be better suited.

DIRECT SOLUTION METHOD

As stated above, the finite difference representation of the heat conduction equation can be cast into a matrix system, so that

$$[A]\{\mathbf{T}\} = \{\mathbf{F}\}$$

where $\{\mathbf{T}\}$ is the vector of unknown temperatures and $[A]$ and $\{\mathbf{F}\}$ are known coefficients and constants. $[A]$ is called the coefficient matrix, and it is a square matrix of order n, where n is the number of unknown nodal temperatures. The solution can be expressed as $\{\mathbf{T}\} = [A]^{-1} \{\mathbf{F}\}$. The problem for the direct solution method basically boils down to calculating $[A]^{-1}$. There are many readily available subroutines for calculating matrix inversions [3]. One of the most commonly used direct solution method is the Gaussian elimination [3, 20]. The method is not efficient if the number of equations to be solved is large, since approximately n^3 multiplications are required to solve n simultaneous equations. Gaussian elimination is efficient when the matrix system is banded with a low bandwidth, particularly for a tridiagonal system.

The basic idea behind Gaussian elimination procedure is as follows. Consider a system of n equations, such as

$$a_{11}T_1 + a_{12}T_2 + \cdots + a_{1n}T_n = b_1$$
$$a_{21}T_1 + a_{22}T_2 + \cdots + a_{2n}T_n = b_2$$
$$a_{n1}T_1 + a_{n2}T_2 + \cdots + a_{nn}T_n = b_n \qquad (3.18)$$

The objective is to transform the equation system into an upper triangular system by eliminating the lower triangular system. This is achieved by multiplying the first equation of the system of n equations by a_{21}/a_{11} and subtracting it from the second equation, thus eliminating T_1 from the second equation. The procedure is continued by multiplying the first equation by a_{31}/a_{11} and subtracting it from the third equation, and so on. This eliminates T_1 from all equations except the first one, which is the pivotal equation for T_1. The procedure is repeated for T_2, T_3, $\ldots$, T_{n-1} to eliminate the coefficients of these unknowns from all of the equations below the diagonal pivot. This results in an upper triangular form of the equation system in Eq. (3.19). The superscript * for a coefficient denotes that the coefficient has been modified from the original system due to the elimination operations performed on these equations.

$$a_{11}T_1 + a_{12}T_2 + \cdots + a_{1n}T_n = b_1$$
$$\cdots + a_{22}^*T_2 + \cdots + a_{2n}^*T_n = b_2^*$$
$$\cdots a_{nn}^*T_n = b_n^*$$

(3.19)

The only unknown remains in the last equation above and so

$$T_n = b_n^*/a_{nn}^*$$

This value of T_n can be used to substitute back in the equation for T_{n-1} and so on, to obtain the solution for the entire system. Although convenient, direct methods such as matrix inversion or Gaussian elimination are not numerically efficient for large nonlinear problems, and it is preferable to use iterative numerical schemes.

ITERATIVE SOLUTION TECHNIQUE

One of the most popular iterative methods for solving the system of linear algebraic equations is the Gauss-Siedel relaxation method. To start the solution, all values of unknown temperatures are initialized to a guessed value. All the unknown values of temperature T are systematically updated until there is no further change within a specified tolerance. At every point (i, j), temperature $T(i,j)$ is updated using the latest value for other temperatures.

This procedure can be illustrated by solving for nodal temperatures T_1, T_2, T_3, and T_4 for the slab problem discussed above. Using Eq. (3.17), the finite difference equation for the four nodes is written using k as the iteration counter as follows:

$$T_1^k = 1/4(50 + T_2^{k-1} + T_3^{k-1} + 100)$$
$$T_2^k = 1/4(T_1^k + 200 + T_4^{k-1} + 100)$$
$$T_3^k = 1/4(50 + T_4^{k-1} + T_1^k + 300)$$
$$T_4^k = 1/4(300 + T_2^k + T_3^k + 200)$$

Start the iterations with a guessed temperature value of 100°C.

k	T_1	T_2	T_3	T_4
0	100	100	100	100
1	87.5	121.875	134.37	189.06
2	101.56	147.655	160.155	201.95
3	114.45	154.1	166.6	205.175
4	117.675	155.71	168.21	205.98
5	118.48	156.115	168.62	206.18
6	118.68	156.22	168.71	206.23

As one can see, the iterations are converging, that is, the temperature change from iteration to iteration is getting smaller and smaller. One has to make a decision when to declare convergence. Usually, when the temperatures do not change within a specified tolerance, convergence is declared.

Sometimes, the rate of convergence of Gauss-Siedel iterations can be enhanced by using overrelaxation. Overrelaxation in simple terms means overextending the iterative guesses. It can be mathematically proven that the overrelaxation parameter ω can take values between 1 and less than 2. The Gauss-Siedel relaxation is $\omega = 1$. Reformulating the equations for our example with the overrelaxation factor, we get

$$T_1^k = \omega/4(50 + T_2^{k-1} + T_3^{k-1} + 100) + (1 - \omega)T_1^{k-1}$$
$$T_2^k = \omega/4(T_1^k + 200 + T_4^{k-1} + 100) + (1 - \omega)T_2^{k-1}$$
$$T_3^k = \omega/4(50 + T_4^{k-1} + T_1^k + 300) + (1 - \omega)T_3^{k-1}$$
$$T_4^k = \omega/4(300 + T_2^k + T_3^k + 200) + (1 - \omega)T_3^{k-1}$$

Iterating with $\omega = 1.25$, we get

k	T_1	T_2	T_3	T_4
0	100	100	100	100
1	84.38	127.34	142.37	211.33
2	105.86	152.73	164.60	199.6
3	116.60	154.44	167.1	206.70
4	117.94	156.02	168.48	205.8
5	118.61	156.13	168.65	206.27
6	118.69	156.24	168.72	206.22

One can observe that with $\omega = 1.25$, node 4 is oscillating toward convergence, while node 1 is slowly converging. Finding an optimum value for the overrelaxation factor ω is not a straightforward procedure. Selecting the overrelaxation factor is an art, and it can speed up convergence for a large system of equations. It is important to point out that one is not limited to using the same value for the overrelaxation factor ω for all the equations; one can choose a different value of ω for each equation. In the preceding example, node 4 is oscillating to convergence, indicating that the value of 1.25 for ω is large. The readers are encouraged to repeat the calculations with an ω value of 1.12 for the equation for T_4^k. Convergence acceleration techniques, such as adaptive relaxation factors used in commercially available industrial computer programs, are discussed in Chapter 6.

The example illustrated above is steady state and has simple Dirichlet-type boundary conditions. More complex boundary conditions and transient problems are introduced in the next chapter on control volume method. The control volume method is physically more intuitive and therefore easier to understand, and easier to apply to boundary conditions.

The iterative methods can be easily programmed for a digital computer. A sample FORTRAN program is given below for the example problem illustrated. Indexed notations are used.

```
            DOUBLE PRECISION T(4,4), TOLD(4,4)
C------Nx = Number of node points in x direction
C------Ny = Number of node points in y direction
C------TOL = Convergence criteria
C------ITMAX = Maximum number of iterations
C------ITER = Iteration number
C------W = relaxation parameter
C------Initialize temperatures, and various parameters

            TOL = 1.e-6
            ITER = 0
            ITMAX = 1000
            DIFF = 0.

            DO 10 I = 2,3
            DO 10 J = 2,3
                    T(I,J) = 100.
                    TOLD(I,J) = 100.
10      CONTINUE

            NX = 4
            NY = 4

C----- Apply boundary conditions

            DO 20 I = 1,4

            T(I,1) = 300.
            T(I,4) = 100.

20      CONTINUE

            DO 30 J = 1,4

            T(1,J) = 50.
            T(4,J) = 200.

30      CONTINUE

C-----Iterate

200     CONTINUE

            DO 40 I = 2,3
            DO 40 J = 2,3

            T(I,J) = 0.25*W*(T(I-1,J) + T(I+1,J) + T(I,J+1), T(I,J-1)
                         + (1-W)*T(I,J)
40      CONTINUE

            ITER = ITER + 1
```

```
        IF(ITER .GT. ITMAX) THEN
        WRITE(6,*)'Max Iterations exceeded'
        GO TO 300
        END IF
```

C------Check Convergence

```
        DO 50 I = 2,3
        DO 50 J = 2,3

        DIFF = DIFF + DABS( T(I,J) - TOLD(I,J))
        TOLD(I,J) = T(I,J)

50      CONTINUE

        IF(DIFF. LT. TOL) THEN
C------Convergence achieved

        DO 60 J =1,4
        JC = NY - J
        WRITE(6,100) (T(I,J), I =1,4)
100     FORMAT (1H, 6x,4E12.4)
60      CONTINUE

        ELSE
        GO TO 200

        END IF

300     CONTINUE
        STOP
        END
```

TRANSIENT HEAT CONDUCTION

Let us again consider the two-dimensinoal system. The governing differential equation for constant, temperature-independent material properties under time-dependent conditions is given by Eq. (2.7) as

$$\frac{\partial T}{\partial t} = \alpha \left[\frac{\partial^2 T}{\partial x^2} + \frac{\partial^2 T}{\partial y^2} \right] \tag{3.20}$$

where $\alpha = k/\rho C_p$ is the thermal diffusivity. We can use the central finite difference equations for spatial derivatives with respect to x and y given by Eqs. (3.9) and (3.12).

Because the problem is time-dependent, it also has to be discretized in time. As it is more intuitive to march ahead in time, a forward difference approximation is used for the time derivative:

$$\left.\frac{\partial T}{\partial t}\right|_{i,j} \approx \frac{T_{i,j}^{n+1} - T_{i,j}^{n}}{\Delta t} \tag{3.21}$$

Superscript n denotes the time level at which the temperature is evaluated. The temperature solution is obtained at each discrete node point, at each discrete time, separated by the time interval Δt. The solution is started at time = 0, where the conditions are known (initial conditions).

Temporal discretization using Eq. (3.21), along with the spatial discretization using Eqs. (3.9) and (3.12), give the complete finite difference discretization of the problem. One still needs to make a decision as to at what time level the temperatures in the finite difference discretization of the spatial derivative are evaluated.

Explicit Method

If the temperatures in the spatial derivative are evaluated at the nth level, this is called the "explicit method." The finite difference representation for Eq. (3.18) using the explicit method for an interior nonboundary node is

$$\frac{1}{\alpha}\frac{T_{i,j}^{n+1} - T_{i,j}^{n}}{\Delta t} = \frac{T_{i+1,j}^{n} - 2T_{i,j}^{n} + T_{i-1,j}^{n}}{\Delta x^2} + \frac{T_{i,j+1}^{n} - 2T_{i,j}^{n} + T_{i,j-1}^{n}}{\Delta y^2} \tag{3.22}$$

If $\Delta x = \Delta y$, the equation reduces to

$$T_{i,j}^{n+1} = \frac{\alpha\Delta t}{\Delta x^2}\left(T_{i+1,j}^{n} + T_{i-1,j}^{n} + T_{i,j+1}^{n} + T_{i,j-1}^{n}\right) + \left(1 - 4\,\frac{\alpha\Delta t}{\Delta x^2}\right)T_{i,j}^{n} \tag{3.23}$$

The term $\alpha\Delta t/\Delta x^2$ is called the Fourier number, Fo.

Equations (3.22) and (3.23) are called explicit because the temperature at a new time level for a node point is calculated from the known temperatures of its neighbors at the previous time level. The temperatures for all node points are known at the starting point, $t = 0$, from the initial conditions. It is therefore straightforward to march through time.

The coefficient of $T_{i,j}^{n}$ in Eq. (3.23), $(1 - 4(\alpha\Delta t/\Delta x^2))$, can become negative. If this coefficient is negative, then increasing the temperature of node point (i, j) results in decreasing the temperatures of its neighbors. This violates the second law of thermodynamics and leads to a physically unrealistic solution. Therefore, to achieve a stable, physically meaningful solution, the time step one can use has to be restricted below a certain limit, based on the value of thermal diffusivity α and discretization Δx.

Implicit Method

The explicit method discussed above is easy to implement; however, it is limited by the restriction placed on the time step. This can lead to extremely small time steps, which in turn lead to an excessive number of computations to march the solution through time. The restriction on time step can be removed by using the implicit method. If the temperatures in the spatial discretization are evaluated at the $(n + 1)$ level (current time level of solution), this leads to a fully implicit solution. The finite difference discretization equation for an interior, nonboundary node using the fully implicit method is

$$\frac{1}{\alpha} \frac{T_{i,j}^{n+1} - T_{i,j}^{n}}{\Delta t} = \frac{T_{i+1,j}^{n+1} - 2T_{i,j}^{n+1} + T_{i-1,j}^{n+1}}{\Delta x^2} + \frac{T_{i,j+1}^{n+1} - 2T_{i,j}^{n+1} + T_{i,j-1}^{n+1}}{\Delta y^2} \quad (3.24)$$

If $\Delta x = \Delta y$, then

$$T_{i,j}^{n} = \left(1 + 4\frac{\alpha \Delta t}{\Delta x^2}\right) T_{i,j}^{n+1} - \frac{\alpha \Delta t}{\Delta x^2} (T_{i+1,j}^{n+1} + T_{i-1,j}^{n+1} + T_{i,j+1}^{n+1} + T_{i,j-1}^{n+1}) \quad (3.25)$$

It can be observed from Eq. (3.25) that the temperature at time level $n + 1$ for $T_{i,j}$ depends on the temperatures of its neighbors at the same time level and therefore is unknown. Consequently, these equations have to be solved simultaneously either by direct or iterative methods, such as Gaussian elimination or Gauss-Siedel iterations. It can also be observed from Eq. (3.25) that the coefficient of $T_{i,j}$ can never become negative. Therefore, there is no restriction on the time step one can take for a stable physically realistic solution. It does not, however, imply that one can take any large time steps and get an accurate solution. A stable and physically meaningful solution does not necessarily mean an accurate solution. Therefore, one still needs to use a prudently chosen time interval.

Chapter 4

Control Volume Method

In the finite difference method, one starts with the governing partial differential equation and uses the Taylor series to represent partial derivatives with finite difference approximations. The partial differential equation (PDE) is accepted as the appropriate form of the physical law governing the heat transfer problem, and mathematical tools are used to develop algebraic approximations to partial derivatives. The physical law, or conservation principle, is used in deriving the differential equation, but then set aside.

In the control volume method, the differential equation governing the problem is examined as it relates to the underlying conservation principle. The conservation principle is applied in the vicinity of each and every discrete grid point or node, i.e., we apply at each and every node point the same conservation principle as is used in deriving the differential equation, but do not take the limit of the control volume to a point. In practice, the control volume method leads quickly to mathematical expressions that have a physical basis and are therefore very versatile when it comes to applying simple or complex boundary conditions. This is because the control volume method keeps the discrete nature of the solution process at every step.

This chapter is intended to give a brief overview of the control volume method. The material presented here is based on reference 4, by Patankar, who provides a detailed description of the method.

ONE-DIMENSIONAL STEADY STATE PROBLEM

Let us visit a one-dimensional steady state heat conduction problem. The governing differential equation with the heat generation term for a one-dimensional region extending from $x = 0$ to $x = L$ is

$$\frac{d}{dx}\left(k\,\frac{dT}{dx}\right) + \dot{q} = 0 \qquad 0 < x < L \qquad (4.1)$$

To write a discrete form of this equation, we partition the one-dimensional problem region, i.e., we put a 1-D mesh on the region of interest. Let's look at a generic point P (see Fig. 4.1). The nomenclature here is taken from Patankar [4].

Next, we integrate the governing equation over the control volume placed on either side of the point of interest, P. The control volume is placed midway

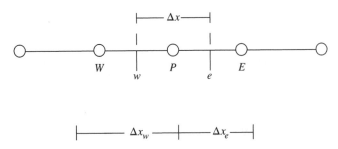

FIGURE 4.1. A generic node point P and the control volume in one dimension.

noindent between the node points. The distance between successive nodes need not be the same. We assume a constant rate of heat generation.

$$\int_w^e \frac{d}{dx}\left(k\,\frac{dT}{dx}\right)dx + \int_w^e \dot{q}\,dx = 0 \qquad (4.2)$$

$$\left(k\,\frac{dT}{dx}\right)_e - \left(k\,\frac{dT}{dx}\right)_w + \dot{q}(x_e - x_w) = 0 \qquad (4.3)$$

To evaluate the temperature gradient dT/dx, we can approximate T as a piecewise linear function between nodes (see Fig. 4.2).

Substituting the piecewise linear behavior between nodes, Eq. (4.3) reduces to

$$\frac{k_e(T_E - T_P)}{\Delta x_e} - \frac{k_w(T_P - T_W)}{\Delta x_w} + \dot{q}\Delta x = 0 \qquad (4.4)$$

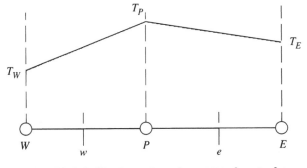

FIGURE 4.2. Temperature is approximated as a piecewise linear function.

Equation (4.4) can be written as

$$a_P T_p = a_E T_E + a_W T_W + b$$

$$a_P = \frac{k_e}{\Delta x_e} + \frac{k_w}{\Delta x_w}$$

$$a_E = \frac{k_e}{\Delta x_e}$$

$$a_W = \frac{k_w}{\Delta x_w}$$

$$b = \dot{q}\Delta x \tag{4.5}$$

If $\Delta x_e = \Delta x_w$ and $k_e = k_w$, then Eq. (4.4) reduces to

$$\frac{2k}{\Delta x} T_P = \frac{k}{\Delta x} T_E + \frac{k}{\Delta x} T_W + \dot{q}\Delta x$$

$$2T_P = T_E + T_W + \frac{\dot{q}}{k} \Delta x^2 \tag{4.6}$$

One can derive the same discretization equation starting from the energy conservation principle applied over the control volume instead of starting from a governing differential equation.

(Energy in) − (Energy out) + (Energy stored) + (Energy generated) = 0

$$\left[\frac{-k_w(T_P - T_W)}{\Delta x_w} \right] - \left[\frac{-k_e(T_E - T_P)}{\Delta x_e} \right] + 0 + \dot{q}\Delta x = 0$$

$$\frac{k_e(T_E - T_P)}{\Delta x_e} - \frac{k_w(T_P - T_W)}{\Delta x_w} + \dot{q}\Delta x = 0$$

The control volume method thus has a foundation in the basic physics of the problem and therefore is more intuitive and easy to use when formulating complex boundary conditions.

BOUNDARY CONDITIONS

To apply the boundary conditions, we have to write a discretization equation for the boundary and adjacent node integrated over half the control volume, as shown in Fig. 4.3.

Dirichlet Boundary Condition (known boundary temperature):

$$a_B T_B = a_I T_I + b$$

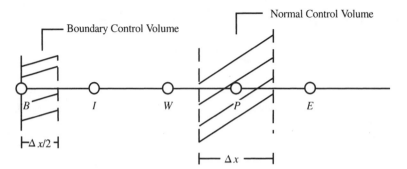

FIGURE 4.3. Boundary control volume with boundary node B and adjacent interior node I.

simply reduces to

$$T_B = T_{\text{known}}$$

or

$$b = T_{\text{known}} \quad \text{and} \quad a_B = 1.0 \tag{4.7}$$

Neumann Boundary Condition (specified heat flux input): The energy balance at the boundary control volume is

$$q_B'' - q_1'' + \dot{q}\Delta x = 0$$

Substituting for the fluxes, we get

$$q_B'' - \frac{k_1(T_B - T_I)}{\Delta x} + \dot{q}\Delta x = 0$$

Recasting results in

$$a_B T_B = a_I T_I + b$$

$$a_I = \frac{k_I}{\Delta x} \qquad b = \dot{q}\Delta x + q_B''$$

$$a_B = a_I \tag{4.8}$$

Mixed Boundary Condition (convection and radiation boundary condition):

$$q_B'' = h(T_f - T_B)$$

where $T_f \equiv$ Fluid temperature
$h \equiv$ Convection heat transfer coefficient

Again, we write the equation in the form

$$a_B T_B = a_I T_I + b$$

$$a_I = \frac{k_I}{\Delta x} \qquad b = q\Delta x + hT_f$$

$$a_B = a_I - q_B'' \Delta x + h \tag{4.9}$$

For radiation, the equation is

$$q_B'' = \epsilon\sigma(T_f^4 - T_B^4)$$

where $\epsilon \equiv$ Emissivity
$\sigma \equiv$ Stefan-Boltzmann constant

Rearranging the equation in the form of a radiation heat transfer coefficient, we have

$$q_B'' = h_r(T_f - T_B)$$

$$h_r = \epsilon\sigma(T_f^2 + T_B^2)(T_f + T_B) \tag{4.10}$$

Replacing h with h_r in Eq. (4.9) results in the equation for the boundary node for this 1-D problem with radiation boundary condition.

Now we have equations for all the nodes. The solution of this system of equations gives the temperature at all the node points. The equations can be written as

$$a_i T_i = b_i T_{i+1} + c_i T_{i-1} + d_i \qquad (i = 1, 2, 3, \cdots, n) \tag{4.11}$$

Note that $c_1 = 0$ and $b_n = 0$. The one-dimensional discrete equations result in a tri-diagonal matrix system, which can be solved very efficiently with a tri-diagonal matrix algorithm, also known as the Thomas algorithm [3, 4].

TRI-DIAGONAL MATRIX ALGORITHM (TDMA)

We know T_1 in terms of T_2 for the $i = 1$ equation, as c_1 is zero. The equation for $i = 2$ is a relationship between T_1, T_2, and T_3 and can be reduced to the relationship between T_2 and T_3 by eliminating T_1 for T_2 from the $i = 1$ equation. We keep substituting forward until T_n is in terms of T_{n+1}. As T_{n+1} does not exist, the last equation gives the value for T_n. Now we can go back and substitute to obtain $T_{n-1}, T_{n-2}, T_{n-3}, \cdots, T_1$. During the forward substitution, we have

$$T_i = P_i T_{i+1} + q_i$$
$$T_{i-1} = p_{i-1} T_i + q_{i-1} \tag{4.12}$$

By substituting Eq. (4.12) into Eq. (4.11), we get

$$a_i T_i = b_i T_{i+1} + c_i(p_{i-1} T_i + q_{i-1}) + d_i \tag{4.13}$$

Rearranging in the form of Eq. (4.12), we get

$$p_i = \frac{b_i}{a_i - c_i P_{i-1}}$$

$$q_i = \frac{d_i + c_i q_{i-1}}{a_i - c_i p_{i-1}} \tag{4.14}$$

These are recurrence relations for p_i and q_i. The TDMA algorithm steps are as follows:

1. Start the process for $i = 1$, $p_1 = b_1/a_1$ and $q_1 = d_1/a_1$
2. At the end, $i = n$ and $p_n = 0$, $T_n = q_n$
3. Use $T_i = P_i T_{i+1} + q_i$ for $i = n - 1, n - 2, \cdots, 1$ to obtain $T_{n-1}, T_{n-2}, \cdots, T_1$

TRANSIENT HEAT CONDUCTION

The governing equation for one-dimensional transient heat conduction is

$$\rho C \frac{\partial T}{\partial t} = \frac{\partial}{\partial x} \left(k \frac{\partial T}{\partial x} \right) \tag{4.15}$$

Let us assume that the density ρ and specific heat C are independent of temperature. (Nonlinear material properties are addressed later, in Chapter 6.)

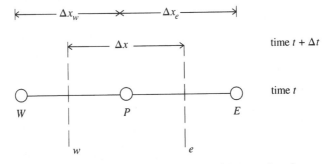

FIGURE 4.4. Typical one-dimensional control volume.

We integrate the differential Eq. (4.15) over the control volume as shown in Fig. 4.4 and over a time interval Δt, from a starting time, t.

$$\rho C \int_w^e \int_t^{t+\Delta t} \frac{\partial T}{\partial t}\, dt\, dx = \int_t^{t+\Delta t} \int_w^e \frac{\partial}{\partial x}\left(k\, \frac{\partial T}{\partial x} \right) dx\, dt \qquad (4.16)$$

Using superscript $n + 1$ for values at time level $t + \Delta t$ and n for values at time t, we get

$$\rho C(T_P^{n+1} - T_P^n)\Delta x = \int_t^{t+\Delta t} \left[\frac{k_e(T_E - T_P)}{\delta x_e} - \frac{k_w(T_P - T_W)}{\delta x_w} \right] dt \qquad (4.17)$$

The integration of the left-hand side of the equation assumes a stepwise behavior over the control volume. This is similar to the forward difference of the time derivative, as discussed in Chapter 3. The integration on the right-hand side can be performed by evaluating the temperatures on the right-hand side either at time level t, $t + \Delta t$, or somewhere in between t and $t + \Delta t$.

If temperatures are evaluated at time level n, i.e., at time t, this results in an explicit equation system, as follows:

$$\rho C(T_P^{n+1} - T_P^n)\Delta x = \left[\frac{k_e}{\Delta x_e}\,(T_E^n - T_P^n) - \frac{k_w}{\Delta x_w}\,(T_P^n - T_W^n) \right] \Delta t \qquad (4.18)$$

Recasting, we get

$$a_P T_P^{n+1} = a_E T_E^n + a_W T_W^n + a_P^n T_P^n$$

$$a_P = \rho C\, \frac{\Delta x}{\Delta t}$$

$$a_P^n = a_P - a_E - a_W$$

$$a_E = \frac{k_e}{\Delta x_e} \qquad a_W = \frac{k_w}{\Delta x_w} \qquad (4.19)$$

Equation (4.19) is an explicit equation for T_P^{n+1} (T_P at time $t + \Delta t$) in terms of known values at time t. The solution therefore is straightforward and can be easily marched in time starting from the initial condition. The caveat in explicit solutions is that the time interval one can take is restricted for a physically realistic solution. As seen in the previous chapter, physically unrealistic solutions result if the discretization coefficients become negative. In Eq. (4.19), a_P^n can become negative if

$$\rho C \, \frac{\Delta x}{\Delta t} < \frac{k_e}{\Delta x_e} + \frac{k_w}{\Delta x_w} \tag{4.20}$$

If $\Delta x_e = \Delta x_W = \Delta x$, then a_P^n becomes negative if

$$\Delta t > \frac{\rho C (\Delta x)^2}{2k} \tag{4.21}$$

Therefore, Δt should be less than $\rho C \Delta x^2 / 2k$. The boundary conditions impose further restrictions on the maximum explicit stable time interval. The explicit stable time step varies from node point to node point.

If temperatures in the right-hand side of Eq. (4.17) are evaluated at the current time, $t + \Delta t$, the result is a fully implicit equation system. The following equation is obtained using the fully implicit scheme:

$$\rho C (T_P^{n+1} - T_P^n)\Delta x = \left[\frac{k_e}{\Delta x_e} (T_E^{n+1} - T_P^{n+1}) - \frac{k_w}{\Delta x_w} (T_P^{n+1} - T_W^{n+1}) \right] \Delta t \tag{4.22}$$

Recasting in the $a_P T_P = a_{\text{neighbor}} T_{\text{neighbor}}$ format, we get

$$a_P T_P^{n+1} = a_E T_E^{n+1} + a_W T'_W + a_P^n T_P^n$$

$$a_P = a_E + a_W + a_P^n$$

$$a_E = \frac{k_e}{\Delta x_e}$$

$$a_W = \frac{k_w}{\Delta x_w}$$

$$a_P^n = \rho C \, \frac{\Delta x}{\Delta t} \tag{4.23}$$

As the temperatures on both sides of Eq. (4.23) are at time $t + \Delta t$, T_P^{n+1} cannot be explicitly calculated from this equation and a system of equations has to be solved to get the temperature solution. That is why the procedure is called fully implicit. Observing the coefficients a_P, a_E, a_W, a_P^n, one can conclude that there is no possibility of getting negative values. Therefore, there is no restriction on the size of the time interval, Δt. Any value for the time interval will give a physically viable solution. However, this does not mean that any value for the time interval will give a numerically accurate or good solution.

If one chooses to evaluate the temperatures in the right-hand side of Eq. (4.17) at a time half way between t and $t + \Delta t$, then this results in what is known as a

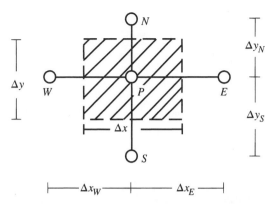

FIGURE 4.5. A generic two-dimensional control volume in a Cartesian system.

Crank-Nicholson scheme or a semi-implicit scheme, and the following equation is obtained:

$$\rho C(T_P^{n+1} - T_P^n)\Delta x = \left[\frac{k_e}{\Delta x_e} \left(\frac{T_E^{n+1} + T_E^n}{2} - \frac{T_P^{n+1} - T_P^n}{2} \right) \right.$$
$$\left. - \frac{k_w}{\Delta x_w} \left(\frac{T_P^{n+1} + T_P^n}{2} - \frac{T_w^{n+1} + T_w^n}{2} \right) \right] \Delta t \quad (4.24)$$

Recasting in our standard format, we get

$$a_P T_P^{n+1} = a_E \frac{T_E^{n+1} + T_E^n}{2} + a_W \frac{T_W^{n+1} + T_W^n}{2} + a_P^n T_P^n$$

$$a_P^n = \rho C \frac{\Delta x}{\Delta t} - \frac{a_E}{2} - \frac{a_W}{2}$$

$$a_E = \frac{k_e}{\Delta x_e} \qquad a_W = \frac{k_w}{\Delta x_w}$$

$$a_P = a_E + a_W + a_P^n \quad (4.25)$$

Note that a_P^n can become negative; however, for this semi-implicit scheme, the oscillations in the solution caused by a negative coefficient die out as the solution marches in time. It is better to select a time interval that will prevent the coefficients from becoming negative.

TWO-DIMENSIONAL PROBLEMS

Figure 4.5 shows a generic two-dimensional control volume. The governing equation without heat generation in a Cartesian coordinate system is

$$\rho C \frac{\partial T}{\partial t} = \frac{\partial}{\partial x}\left(k_x \frac{\partial T}{\partial x}\right) + \frac{\partial}{\partial y}\left(k_y \frac{\partial T}{\partial y}\right) \tag{4.26}$$

Integrating over the control volume and over time, we get

$$\rho C \int_{cv}\int_{t}^{t+\Delta t} \frac{\partial T}{\partial t}\, dt\, d\text{vol.} = \int_{t}^{t+\Delta t} \int_{cv}\left(\frac{\partial}{\partial x}\left(k_x \frac{\partial T}{\partial x}\right) + \frac{\partial}{\partial y}\left(k_y \frac{\partial T}{\partial y}\right)\right) d\text{vol.}\, dt \tag{4.27}$$

or

$$\rho C \Delta x \Delta y.1(T_P^{n+1} - T_P^n) = \int_{t}^{t+\Delta t}\left[\frac{k_e \Delta y(T_E - T_P)}{\Delta x_e} - \frac{k_w \Delta y(T_P - T_W)}{\Delta x_w}\right.$$
$$\left. + \frac{k_N \Delta x(T_N - T_P)}{\Delta y_n} - \frac{k_s \Delta x(T_P - T_S)}{\Delta y_s}\right] dt \tag{4.28}$$

Similar to the 1-D problem, we can choose an explicit, fully implicit, or Crank-Nicholson scheme. For a fully implicit scheme, we get the following equations:

$$a_P T_P = a_E T_E + a_W T_W + a_N T_N + a_S T_S + b$$

$$a_E = \frac{k_e \Delta x \Delta y}{\Delta x_e}$$

$$a_W = \frac{k_w \Delta x \Delta y}{\Delta x_w}$$

$$a_N = \frac{k_n \Delta x \Delta y}{\Delta x_n}$$

$$a_S = \frac{k_s \Delta x \Delta y}{\Delta x_s}$$

$$a_P^n = \frac{\rho C \Delta x \Delta y}{\Delta t}$$

$$b = a_P^n T_P^n$$

$$a_P = a_E + a_W + a_N + a_S + a_P^n \tag{4.29}$$

This formulation can be easily extended to three-dimensional problems by just adding two more neighboring node points. The equations for 2-D and 3-D heat conduction problems are not tri-diagonal in nature. Two-dimensional problems lead to a pent-diagonal system, while three-dimensional problems result in a hept-diagonal system. There are routines such as TDMA for pent-diagonal systems; however, they are not as efficient, and usually iterative techniques are used to solve these systems of equations. Gauss-Siedel, discussed in the previous chapter, is one of the most popular iterative techniques. If the geometry of the physical domain is simple enough to be represented by mesh lines that run parallel to the coordinate axes, then the powerful one-dimensional TDMA algorithm can be used in two or three orthogonal directions. The readers are referred to reference 4 for details of this procedure. A more detailed description of the control volume method as it is applied to conduction problems as well as fluid flow (CFD) problems can also be found in [4].

In most industrial problems, the shapes and sizes encountered for the physical domain are highly irregular and the models are built with off-the-shelf commercial CAD/CAE packages that serve a multitude of analytical needs, such as structural, thermal, fluid, or electromagnetic analyses, among others. Most of these CAD/CAE packages come with built-in meshing algorithms that generate finite element meshes. One can use these meshes with the control volume method, but the implementation is not straightforward unless the geometry lines up with the coordinate axes. Sometimes the irregular geometry is enclosed in meshes that are parallel to the coordinate axes and the curvature is modeled as a series of small stair steps. The finite element method is more suited to problems with irregular or complex geometries, and it is discussed in the next chapter.

Chapter 5

Finite Element Method

The finite element method (FEM) uses piecewise or regional approximations to replace differential equations. The finite difference method (FDM), on the other hand, provides point-by-point approximations. The finite difference method is easy to implement in computer programs; however, it suffers from limitations in its ability to negotiate complex geometries. In contrast, the finite element method can be applied to any geometry. Because denoting the complex physical domains as finite elements in mathematical terms leads to programming complexities, the finite element method is not as straightforward to implement as the finite difference method. Many commercially available, easy-to-use finite element preprocessors make creating the geometry of the physical domain and finite element meshes less painstaking.

The finite element method has three major steps, as follows:

1. The physical domain is divided into many smaller subdomains, called elements. The whole region is viewed as a collection of these small regions, or elements.
2. For each element the solution of the equation (e.g., temperature distribution) is approximated by interpolation polynomials.
3. Since the region is viewed as a collection of elements, these individual approximations are put together, or assembled, to obtain the finite element representation of the region.

As the solution is approximated by interpolation polynomials on each element, to obtain a continuous solution, one has to impose continuity of solution, and maybe its derivative, across element boundaries.

There is a wealth of reference material available on finite element method in general [5, 6] and finite element heat transfer in particular [7, 8]. The objective of this chapter is to briefly introduce the fundamental concepts of finite element heat transfer for conduction problems (convection and radiation are considered only as boundary conditions) and to introduce finite element formulations for 1-D, 2-D, 3-D elements. These formulations are used in an innovative way to generate an equivalent conductance-capacitor network, which then can be solved by network solution techniques, as is discussed in detail in the next chapter.

GALERKIN METHOD

The steady state heat conduction equation with heat generation in one dimension is

$$k \frac{d^2T}{dx^2} + \dot{q} = 0 \tag{5.1}$$

To obtain a finite element solution for Eq. (5.1), one first has to subdivide the region into a number of finite elements, say n. This is illustrated in Fig. 5.1. The one-dimensional physical domain between $x = 0$ and $x = L$ is subdivided into n finite elements $e_1, e_2, \cdots, e_n$. The temperature distribution $T(x)$ is approximated over each of these n elements by using known polynomial functions $\phi_j(x)$ and constants T_j. Therefore,

$$T(x) \approx T_1\phi_1(x) + T_2\phi_2(x) + T_3\phi_3(x) + \cdots + T_{n+1}\phi_{n+1}(x)$$

$$T(x) \cong \sum_{j=1}^{n+1} T_j\phi_j \tag{5.2}$$

ϕ is called a shape function or a basis function. If we substitute Eq. (5.2) into the governing differential equation, there will be some nonzero remainder called residue R. Equation (5.2) is only an approximate solution. Consequently,

$$R = k\,\frac{d^2T}{dx^2} + \dot{q}$$

The basic idea behind the method of weighted residuals, such as the Galerkin method, is not to make the residue vanish at every point (which is impossible), but to make it vanish at the node points of the mesh. The residue, or the error, in solution is made to vanish in some weighted sense, i.e., we multiply the residue function by a weighting function $w(x)$ and make the integral of the product of residue and weighting function go to zero:

$$\int_0^L w(x)R(x)dx = 0 \tag{5.3}$$

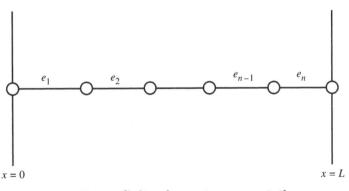

$x = 0$ $x = L$

**FIGURE 5.1. A finite element representation
of a one-dimensional domain.**

In the Galerkin weighted residual method, the shape functions, ϕ_j, are used as the weighing functions, $w_j(x)$:

$$\int_0^L \phi(x) \left[-k \frac{d^2T}{dx^2} - \dot{q} \right] dx = 0 \qquad (5.4)$$

Integrating by parts, we get

$$\int_0^L k \frac{dT}{dx} \frac{d\phi}{dx} - k\phi \left. \frac{dT}{dx} \right|_0^L - \int_0^L \phi(x)\dot{q} \, dx = 0 \qquad (5.5)$$

Equation (5.5) is sometimes referred to as the "weak statement" of the governing differential equation because it has only a first derivative term while the governing differential equation has a second-order derivative of temperature. The requirement on solution function T has been weakened since the temperature solution has to be differentiable only once.

BOUNDARY CONDITIONS

If the boundary condition is of Dirichlet type, i.e., has a specified temperature, the shape function at that boundary is set to zero. In the above example, if temperature at $x = 0$ is known, i.e., $T(x = 0) = T_s$, then the flux term at $x = 0$ is set to zero, by setting $\phi(0) = 0$.

$$-k\phi \left. \frac{dT}{dx} \right|_{x=0} = 0$$

If the boundary condition is of Neumann type (specified heat flux) or mixed (convection or radiation), the second term in Eq. (5.5) is replaced by the appropriate flux and shape function. Therefore, if heat flux is specified at the surface $x = L$, then

$$-k \left. \frac{dT}{dx} \right|_{x=L} = q''$$

$$-k\phi \left. \frac{dT}{dx} \right|_{x=L} = \phi(L)q'' \qquad (5.6)$$

If a convection boundary condition is applied at $x = L$, then

$$-k \left. \frac{dT}{dx} \right|_{x=L} = h(T - T_\infty)$$

Therefore,

$$-k \left. \frac{dT}{dx} \right|_{x=L} = \phi(L)h(T - T_\infty) \tag{5.7}$$

The term $\phi(L)$ is usually set to 1, i.e., the shape function (ϕ) at the boundary where flux or convection is specified is set to 1. The shape functions therefore never appear explicitly in the formulations at the boundaries.

The temperature solution is approximated by

$$T(x) = \sum_{j=1}^{n+1} T_j \phi_j(x)$$

Substituting in Eq. (5.5), we get

$$\sum_{j=1}^{n+1} T_j \left[\int_0^L k \frac{d\phi_i}{dx} \frac{d\phi_j}{dx} \, dx \right] - \int_0^L \phi_i \dot{q} \, dx - k\phi_i \left. \frac{dT}{dx} \right|_{x=0} = 0 \tag{5.8}$$

Once the shape functions are chosen, one can compute the integrals, and Eq. (5.7) is transformed into algebraic simultaneous equations for T_j, i.e., temperatures at the discrete node points.

ONE-DIMENSIONAL ELEMENT SHAPE FUNCTIONS

The shape functions can be chosen to be any mathematical function; usually polynomial functions are selected. One can choose a polynomial function that is applicable throughout the region; however, it is convenient and therefore normal practice to use a set of polynomial functions that are applicable only over a certain region and vanish everywhere else.

As shown in Fig. 5.2, the one-dimensional region of length L is partitioned into n elements and $n + 1$ nodes, with $x_1 = 0$ and $x_{n+1} = L$. Each node has a shape function associated with it, i.e., node i has shape function ϕ_i associated with it. These one-dimensional functions are sometimes called "hat functions" because of their shape: they peak at the node point (i) and for linear elements go down to zero at nodes $i - 1$ and $i + 1$ on either side of node i. Therefore, $\phi_i(x_i) = 1$, and

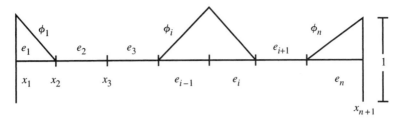

FIGURE 5.2. One-dimensional shape functions (hat functions).

if $i \neq j$, then $\phi_i(x_j) = 0$, so that

$$\phi_i(x) = \frac{x - x_{i-1}}{\Delta x_{i-1}} \qquad x_{i-1} \leq x \leq x_i \qquad \text{and} \qquad i = 1, 2, \ldots, n$$

$$= \frac{x_{i+1} - x}{\Delta x_i} \qquad x_i \leq x \leq x_{i+1} \qquad \text{and} \qquad \Delta x_i = x_{i+1} - x_i$$

$$= 0 \qquad \text{elsewhere} \tag{5.9}$$

The temperature distribution is once again approximated as

$$T(x) = \phi_1(x)T_1 + \phi_2(x)T_2 + \cdots + \phi_{n+1}(x)T_{n+1} = \sum_{j=1}^{n+1} \phi_j T_j \tag{5.10}$$

One can construct shape functions that are quadratic, cubic, etc. The readers are referred to advanced texts [7, 8] for their formulations. Higher order elements can be computationally intensive, especially for multidimensional problems.

To illustrate the application of the method, let us solve the steady state, one-dimensional conduction problem with internal heat generation discussed earlier with two finite elements and three nodes. We use the weak form given in Eq. (5.5) with the hatlike shape functions. Let the finite element be of equal length $L/2$. Writing the Galerkin formulation for each element, we have

$$\int_0^{L/2} \left[k \frac{d\phi_i}{dx} \sum_{j=1}^{3} T_j \frac{d\phi_j}{dx} \right] - \left[\int_0^{L/2} \phi_i \dot{q} \, dx \right] - \left[k\phi_i \frac{dT}{dx} \right] \Bigg|_{x=0}^{L/2}$$

$$+ \int_{L/2}^{L} \left[k \frac{d\phi_i}{dx} \sum_{j=1}^{3} T_j \frac{d\phi_j}{dx} \right] - \left[\int_{L/2}^{L} \phi_i \dot{q} \, dx \right] - \left[\phi_i k \frac{dT}{dx} \right] \Bigg|_{x=L/2}^{L} = 0 \tag{5.11}$$

The terms for integrals from 0 to $L/2$ are for element 1, and the terms for integrals from $L/2$ to L are for element 2. For element 1 over 0–$L/2$, only the shape functions ϕ_1 and ϕ_2 exist, and ϕ_3 is zero. Expanding Eq. (5.11) in matrix form, we get

$$\int_0^{L/2} \left[k \begin{bmatrix} \dfrac{d\phi_1}{dx} \\ \dfrac{d\phi_2}{dx} \end{bmatrix} \begin{bmatrix} \dfrac{d\phi_1}{dx} & \dfrac{d\phi_2}{dx} \end{bmatrix} \right] \begin{bmatrix} T_1 \\ T_2 \end{bmatrix} dx$$

$$= \int_0^{L/2} \begin{bmatrix} \phi_1 \\ \phi_2 \end{bmatrix} \dot{q} dx + \begin{bmatrix} \phi_1 \\ \phi_2 \end{bmatrix} \left[-k \dfrac{dT}{dx} \right]\Bigg|_0^{L/2} = \begin{bmatrix} 0 \\ 0 \end{bmatrix} \tag{5.12}$$

Using the linear shape functions

$$\frac{d\phi_1}{dx} = \frac{-2}{L} \qquad \frac{d\phi_2}{dx} = \frac{2}{L} \qquad \phi_1 = 1 - \frac{x}{L/2} \qquad \phi_2 = \frac{x}{L/2} \tag{5.13}$$

the element formulation for element 1 reduces to

$$\int_0^{L/2} k \begin{bmatrix} \dfrac{4}{L^2} & -\dfrac{4}{L^2} \\ -\dfrac{4}{L^2} & \dfrac{4}{L^2} \end{bmatrix} \begin{bmatrix} T_1 \\ T_2 \end{bmatrix} dx - \int_0^{L/2} \begin{bmatrix} 1 - \dfrac{x}{L/2} \\ \dfrac{x}{L/2} \end{bmatrix} \dot{q} dx$$

$$+ \begin{bmatrix} \left(1 - \dfrac{L/2}{L/2}\right)\left(-k\dfrac{dT}{dx}\right)\Big|_{x=L/2} & -(1)\left(-k\dfrac{dT}{dx}\right)\Big|_{x=0} \\ \dfrac{L/2}{L/2}\left(-k\dfrac{dT}{dx}\right)\Big|_{x=L/2} & -\left(\dfrac{0}{L/2}\right)\left(-k\dfrac{dT}{dx}\right)\Big|_{x=0} \end{bmatrix} = \begin{bmatrix} 0 \\ 0 \end{bmatrix}$$

$$\tag{5.14}$$

Therefore,

$$\frac{k}{L/2} \begin{bmatrix} 1 & -1 \\ -1 & 1 \end{bmatrix} \begin{bmatrix} T_1 \\ T_2 \end{bmatrix} = \frac{\dot{q}L}{4} \begin{bmatrix} 1 \\ 1 \end{bmatrix} + \begin{bmatrix} -k\dfrac{dT}{dx}\Big|_{x=0} \\ k\dfrac{dT}{dx}\Big|_{x=L/2} \end{bmatrix} \tag{5.15}$$

For the second element, we have

$$\int_{L/2}^{L} k \begin{bmatrix} \dfrac{d\phi_2}{dx} \\[2ex] \dfrac{d\phi_3}{dx} \end{bmatrix} \begin{bmatrix} \dfrac{d\phi_2}{dx} & \dfrac{d\phi_3}{dx} \end{bmatrix} \begin{bmatrix} T_2 \\ T_3 \end{bmatrix} dx - \int_{L/2}^{L} \begin{bmatrix} \phi_2 \\ \phi_3 \end{bmatrix} \dot{q}\, dx$$

$$+ \begin{bmatrix} \phi_2 \\ \phi_3 \end{bmatrix} \left(-k \dfrac{dT}{dx} \right)\Bigg|_{x=L/2}^{x=L} = \begin{bmatrix} 0 \\ 0 \end{bmatrix} \tag{5.16}$$

Again, using linear shape functions

$$\dfrac{d\phi_2}{dx} = \dfrac{2}{L} \qquad \dfrac{d\phi_3}{dx} = \dfrac{-2}{L} \qquad \phi_2 = \dfrac{x}{L/2} \qquad \phi_3 = 1 - \dfrac{x}{L/2} \tag{5.17}$$

the element formulation for element 2 reduces to

$$\int_{L/2}^{L} k \begin{bmatrix} \dfrac{4}{L^2} & \dfrac{-4}{L^2} \\[2ex] \dfrac{-4}{L^2} & \dfrac{4}{L^2} \end{bmatrix} \begin{bmatrix} T_2 \\ T_3 \end{bmatrix} dx - \int_{L/2}^{L} \begin{bmatrix} \dfrac{2}{L}(L-x) \\[2ex] \dfrac{x}{L/2} - 1 \end{bmatrix} \dot{q}\, dx$$

$$+ \begin{bmatrix} \dfrac{2}{L}(L-L)\left(-k\dfrac{dT}{dx}\right)\Big|_{x=L} - \left(-k\dfrac{dT}{dx}\right)\Big|_{x=L/2} \dfrac{2}{L}\left(L-\dfrac{L}{2}\right) \\[3ex] \left(-k\dfrac{dT}{dx}\right)\Big|_{x=L} - \dfrac{1}{L}(L-L) \end{bmatrix} = \begin{bmatrix} 0 \\ 0 \end{bmatrix}$$

$$\tag{5.18}$$

Therefore,

$$\dfrac{2k}{L} \begin{bmatrix} 1 & -1 \\ -1 & 1 \end{bmatrix} \begin{bmatrix} T_2 \\ T_3 \end{bmatrix} = \dfrac{\dot{q}L}{4} \begin{bmatrix} 1 \\ 1 \end{bmatrix} + \begin{bmatrix} -k\dfrac{dT}{dx}\Big|_{x=L/2} \\[2ex] k\dfrac{dT}{dx}\Big|_{x=L} \end{bmatrix} = \begin{bmatrix} 0 \\ 0 \end{bmatrix} \tag{5.19}$$

The next step in the finite element solution is assembling the individual element matrix equations into a global matrix:

$$
\begin{bmatrix}
\dfrac{2k}{L} & \dfrac{-2k}{L} & \\[2mm]
\dfrac{-2k}{L} & \dfrac{2k}{L}+\dfrac{2k}{L} & \dfrac{-2k}{L} \\[2mm]
& \dfrac{-2k}{L} & \dfrac{2k}{L}
\end{bmatrix}
\begin{bmatrix} T_1 \\ T_2 \\ T_3 \end{bmatrix}
=
\begin{bmatrix}
\dfrac{\dot{q}L}{4} + -k\ \dfrac{dT}{dx}\Big|_{x=0} \\[3mm]
\dfrac{2\dot{q}L}{4} + k\ \dfrac{dT}{dx}\Big|_{x=L/2} - k\ \dfrac{dT}{dx}\Big|_{x=L/2} \\[3mm]
\dfrac{\dot{q}L}{4} + k\ \dfrac{dT}{dx}\Big|_{x=L}
\end{bmatrix}
$$

$$(5.20)$$

These assembled equations have to be solved after applying the boundary conditions with various matrix solution methods such as Gaussian elimination. One can observe that the flux term $(-k\,dT/dx)$ for the interior node points cancels out. This is to be expected from the physics of heat conduction, which dictates that the heat flux has to be continuous.

TRANSIENT HEAT CONDUCTION IN ONE DIMENSION

Let us look at a simple one-dimensional transient heat conduction problem:

$$\frac{\partial T}{\partial t} = \frac{k}{\rho C_P}\frac{\partial^2 T}{\partial x^2} \qquad \frac{k}{\rho C_P} \equiv \alpha \equiv \text{Thermal diffusivity} \qquad (5.21)$$

Since temperature is a function of space as well as time, to completely define the problem mathematically, one has to specify initial conditions, that is, the known temperature distribution at the starting time $t = 0$. The boundary conditions can also be functions of time. The weighted residual statement for the governing transient equation is

$$\int_0^L w(x)\left[\frac{\partial T}{\partial t} - \alpha\,\frac{\partial^2 T}{\partial x^2}\right] \qquad (5.22)$$

The weighting functions are functions of space only. The weak statement for the second-order space derivative is the same as discussed for the steady state solution. Assume the temperature distribution to be

$$T(x,t) = \sum_{j=1}^{n+1} \phi_j T_j$$

$$\frac{\partial T}{\partial t} = \sum_{j=1}^{n+1} \phi_j\,\frac{\partial T_j}{\partial t} \qquad (5.23)$$

Then, for the transient part of the equation, using the weighting function as the shape function, ϕ, we get

$$\int_0^L \left[\phi_i \sum_{j=1}^{n+1} \phi_j \frac{\partial T_j}{\partial t} \right] dx = \int_0^L \phi_i \sum_{j=1}^{n+1} \phi_j dx \frac{\partial T_j}{\partial t} \tag{5.24}$$

The term

$$\int_0^L \left(\phi_i \sum_{j=1}^{n+1} \phi_j \right) dx$$

is known as the mass matrix. The integral physically represents the area or volume of the element at node i connected to all nodes j and is commonly known as the nodal subarea or volume.

The time derivative of the temperature is usually replaced by a simple forward difference:

$$\frac{\partial T}{\partial t} = \frac{T^{n+1} - T^n}{\Delta t}$$

where

$$T^{n+1} = T(x, t + \Delta t) \quad \text{and} \quad T^n = T(x, t)$$

or

$$T^{n+1} = T(x, t_{n+1}) \quad \text{and} \quad T^n = T(x, t_n) \tag{5.25}$$

The solution at t_n is known and is used to progress the solution to t_{n+1}. Gathering the transient and steady state part of the weak statements of the governing equation

$$\left[\int_0^L \phi_i \phi_j dx \right] \frac{\partial T_i}{\partial t} + \left[\int_0^L \alpha \frac{\partial \phi_i}{\partial x} \frac{\partial \phi_j}{\partial x} dx \right] T_j - \alpha \phi_i \frac{\partial T}{\partial x} \Big|_{x=0}^{x=L} = 0 \tag{5.26}$$

Substituting for the time derivative as discussed above, we get

$$\left[\int_0^L \phi_i \phi_j dx \right] \frac{T_j^{n+1} - T_j^n}{\Delta t} + \left[\int_0^L \alpha \frac{\partial \phi_i}{\partial x} \frac{\partial \phi_j}{\partial x} dx \right] T_j - \alpha \phi_i \frac{\partial T}{\partial x} \Big|_{x=0}^{x=L} = 0 \tag{5.27}$$

At this point, one has to make a decision at what time level to evaluate the temperature and the fluxes for the second and third term. We can use a technique similar to the one discussed in the previous chapters on finite difference method and control volume method, as follows:

$$T = \theta T^{n+1} + (1 - \theta)T^n \qquad 0 \le \theta \le 1 \qquad (5.28)$$

Substituting and gathering terms for T^{n+1} and T^n, we get

$$\left[\int_0^L \phi_i \phi_j dx + \alpha \Delta t \theta \int_0^L \frac{\partial \phi_i}{\partial x} \frac{\partial \phi_j}{\partial x} \, dx \right] T_j^{n+1} + \theta \Delta t \left[\phi_i - k \frac{\partial T^{n+1}}{\partial x} \right]_{x=0}^{x=L}$$

$$+ (1 - \theta)\Delta t \left[\phi_i - k \frac{\partial T^n}{\partial x} \right]_{x=0}^{x=L} = \left[\int_0^L \phi_i \phi_j dx + \alpha \Delta t(\theta - 1) \int_0^L \frac{\partial \phi_i}{\partial x} \frac{\partial \phi_j}{\partial x} \, dx \right] T_j^n$$

$$(5.29)$$

Note that $\theta = 1$ leads to a fully implicit scheme, while $\theta = 0$ results in an explicit formulation, and $\theta = 1/2$ is the Crank-Nicholson method. As discussed in the previous chapters, the explicit method imposes a restriction on the increment that can be used for a stable, physically viable solution. Implicit and Crank-Nicholson methods do not impose a restriction on the possible time interval. This, however, does not mean that one can use any time step one desires and still get a good, accurate solution. The solution with the fully implicit method (i.e., with $\theta = 1$) will be stable, meaning that it will be physically realistic and won't violate thermodynamic laws regardless of the time step used. Usually, in commercial finite element software, one starts with a small initial time step, say 0.001, and adjusts it based on convergence history or condition. The readers are referred to *MSC/NASTRAN*[TM] *Thermal Analysis Handbook* [9] for details on the automatic time stepping algorithm.

For transient problems with nonlinear material properties and boundary conditions, the element matrix evaluation, assembly, and solution have to be performed every iteration within each time interval and there are many such time steps. That is why highly nonlinear transient problems are computationally intensive with the finite element method.

TWO-DIMENSIONAL ELEMENTS

In two dimensions, the region is discretized by using either triangular or quadrilateral discrete elements. Let us study the element formulations of triangular (tri) and quadrilateral (quad) elements.

Triangular Element

Figure 5.3 shows a typical linear triangular element. The shape functions for a linear triangular element are defined as

$$\phi_1(x, y) = \frac{1}{2A} \left[(x_2 y_3 - x_3 y_2) + (y_2 - y_3)x + (x_3 - x_2)y \right]$$

$$\phi_2(x, y) = \frac{1}{2A} \left[(x_3 y_1 - x_1 y_3) + (y_3 - y_1)x + (x_1 - x_3)y \right]$$

$$\phi_3(x, y) = \frac{1}{2A} \left[(x_1 y_2 - x_2 y_1) + (y_1 - y_2)x + (x_2 - x_1)y \right] \quad (5.30)$$

A is the area of the element. The shape function ϕ_1 vanishes (is equal to zero) at nodes 2 and 3. Similarly, ϕ_2 and ϕ_3 vanish at nodes 1 and 3 and nodes 1 and 2, respectively. To obtain ϕ_1 at node 1, we substitute $x = x_1$ and $y = y_1$ in Eq. (5.30), so that

$$\phi_1(x_1, y_1) = \frac{1}{2A} \left[(x_2 y_3 - x_3 y_2) + (y_2 x_1 - y_3 x_1) + (x_3 y_1 - x_2 y_1) \right]$$

$$= \frac{2A}{2A} = 1$$

It is common practice to use an area coordinate system when formulating the element matrix for a triangular element.

$$L_1 = \frac{A_1}{A} \qquad L_2 = \frac{A_2}{A} \qquad L_3 = \frac{A_3}{A} \quad (5.31)$$

$A_1, A_2,$ and A_3 are partial areas that are defined, as shown in Fig. 5.4, by connecting the centroid of the element to corner nodes. The lines of constant L_1 value run parallel to the side opposite node 1. The shape functions for the linear triangular

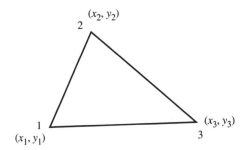

FIGURE 5.3. Typical linear triangular element.

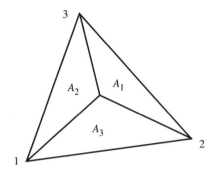

FIGURE 5.4. Area coordinate system.

element therefore can be expressed as

$$\phi_1 = L_1 \qquad \phi_2 = L_2 \qquad \phi_3 = L_3 \tag{5.32}$$

Transient Heat Conduction with Triangular Element

Let us evaluate the element matrix for the linear triangular element. The governing equation for transient heat conduction with internal heat generation is

$$\alpha \left[\frac{\partial^2 T}{\partial x^2} + \frac{\partial^2 T}{\partial y^2} \right] + \frac{\dot{q}}{\rho C_p} = \frac{\partial T}{\partial t} \tag{5.33}$$

The boundary conditions could be of Dirichlet type

$$T = T_B$$

or Neumann type

$$-k \frac{\partial T}{\partial n} = q''$$

or mixed

$$-k \frac{\partial T}{\partial n} = h(T - T_\infty)$$

Multiplying the governing equation by a weighting function w, integrating by parts, and making use of Green's theorem in a plane, we get

$$\int_{Area} w \left(\frac{\partial T}{\partial t} - \alpha \frac{\partial^2 T}{\partial x^2} - \alpha \frac{\partial^2 T}{\partial y^2} - \frac{\dot{q}}{\rho C_p} \right) dA$$

$$\int_{Area} w \frac{\partial T}{\partial t} + \int_{Area} \left(\alpha \left[\frac{\partial w}{\partial x} \frac{\partial T}{\partial x} + \frac{\partial w}{\partial y} \frac{\partial T}{\partial y} \right] - w \frac{\dot{q}}{\rho C_p} \right) dA$$

$$+ \int_{Boundary} w \left(-\alpha \frac{\partial T}{\partial n} \right) d\text{Boundary} = 0 \qquad (5.34)$$

Approximating the temperature field with shape functions, we have

$$T(x,y) = \sum_{j=1}^{\#Nodes} \phi_j(x,y) T_j \qquad (5.35)$$

and using Galerkin procedure, i.e., weighting function $w_i = \phi_i$, we get

$$\left[\int_{Area} \phi_i \phi_j dA \right] \frac{\partial T_i}{\partial t} + \sum_{j=1}^{\#Nodes} \left[\int_{Area} \alpha \left[\frac{\partial \phi_i}{\partial x} \frac{\partial \phi_j}{\partial x} + \frac{\partial \phi_i}{\partial y} \frac{\partial \phi_j}{\partial y} \right] dA \right] T_j$$

$$- \int_{Area} \phi_i \frac{\dot{q}}{\rho C_p} dA + \int_{Boundary} \phi_i \left(-\alpha \frac{\partial T}{\partial n} \right) d\text{Boundary} = 0 \qquad (5.36)$$

One can see from the last term of the above equation that the flux boundary condition falls out naturally from the finite element formulation, which is why it is sometimes referred to as a natural boundary condition. Let us assume that flux at one of the boundaries is known and is q''. The index i also goes from 1 to number of nodes (# Nodes) in the model. The first term in Eq. (5.36) is known as the mass matrix, the second term is the conductivity matrix, and the last two terms are part of the right-hand side or the load vector. After integrations are carried out, the conductivity matrix is given by

$$k_{ij} = \frac{k}{4A} \begin{bmatrix} b_1^2 + c_1^2 & b_2 b_1 + c_2 c_1 & b_3 b_1 + c_3 c_1 \\ b_2 b_1 + c_2 c_1 & b_2^2 + c_2^2 & b_2 b_3 + c_2 c_3 \\ b_3 b_1 + c_3 c_1 & b_2 b_3 + c_2 c_3 & b_3^2 + c_3^2 \end{bmatrix} \qquad (5.37)$$

where the b's and c's are given by

$$b_1 = y_2 - y_3 \qquad b_2 = y_3 - y_1 \qquad b_3 = y_1 - y_2$$
$$c_1 = x_3 - x_2 \qquad c_2 = x_1 - x_3 \qquad c_3 = x_2 - x_1 \qquad (5.38)$$

The mass matrix is

$$\int_A \phi_i\phi_j dA = \int_A \begin{bmatrix} L_1 \\ L_2 \\ L_3 \end{bmatrix} [L_1 \quad L_2 \quad L_3] dA$$

$$= \int_A \begin{bmatrix} L_1^2 & L_1L_2 & L_1L_3 \\ L_2L_1 & L_2^2 & L_2L_3 \\ L_3L_1 & L_3L_2 & L_3^2 \end{bmatrix} dA \qquad (5.39)$$

When we use the integration formulas described earlier for the area coordinates L, the mass matrix becomes

$$M = \frac{A}{12} \begin{bmatrix} 2 & 1 & 1 \\ 1 & 2 & 1 \\ 1 & 1 & 2 \end{bmatrix} \qquad (5.40)$$

where A = Area of the element

To evaluate the load vector integrals, we use the integration formulas for the area coordinates L. The line integral is

$$\int_0^L L_1^a L_2^b dx = \frac{a!\,b!}{(1+a+b)!} L \qquad (5.41)$$

where a and b are non-negative integers and L is the length of the line segment. The area integral is

$$\int_{Area} L_1^a L_2^b L_3^c dA = \frac{a!\,b!\,c!}{(2+a+b+c)!} 2A \qquad (5.42)$$

Therefore,

$$\int_{Area} \dot{q} \begin{bmatrix} \phi_1 \\ \phi_2 \\ \phi_3 \end{bmatrix} dA = \int_A \dot{q} \begin{bmatrix} L_1^1 & L_2^0 & L_3^0 \\ L_1^0 & L_2^1 & L_3^0 \\ L_1^0 & L_2^0 & L_3^1 \end{bmatrix} dA = \frac{\dot{q}A}{3} \begin{bmatrix} 1 \\ 1 \\ 1 \end{bmatrix} \qquad (5.43)$$

The line integral depends on which line segment (i.e., connecting which two nodes) is used. If the line is connecting nodes 1 and 2, then

$$\int_B \dot{q} \begin{bmatrix} L_1^1 L_2^0 \\ L_1^0 L_2^1 \\ 0 \end{bmatrix} d\text{Boundary} = \frac{\dot{q} L_{1-2}}{2} \begin{bmatrix} 1 \\ 1 \\ 0 \end{bmatrix} \tag{5.44}$$

The matrix equation for the element is

$$\underline{M^e T + K^e T = F^e} \tag{5.45}$$

where M^e is the element mass matrix, K^e is the element conductivity matrix, and F^e is the element load vector. One can approximate the time derivative of temperature as

$$\dot{T} = \frac{T^{n+1} - T^n}{\Delta t}$$

where n is the time level, and then use the θ method discussed earlier for 1-D formulation. To obtain a fully explicit scheme, the mass matrix is lumped, that is, all the terms in a row of the mass matrix are added and the sum is put on the diagonal. For each element in the model, we will have an element conductivity matrix, a mass matrix, and a load vector that is assembled in the global model matrix and is solved for temperature distribution at nodes.

Example 5.1
Find the conductivity or stiffness matrix for the triangular element shown below.

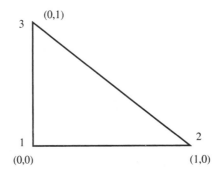

Using Eqs. (5.37) and (5.38), we have

$$k_{ij} = \frac{k}{4A} \begin{bmatrix} b_1^2 + c_1^2 & b_2b_1 + c_2c_1 & b_3b_1 + c_3c_1 \\ b_2b_1 + c_2c_1 & b_2^2 + c_2^2 & b_2b_3 + c_2c_3 \\ b_3b_1 + c_3c_1 & b_2b_3 + c_2c_3 & b_3^2 + c_3^2 \end{bmatrix}$$

$$b_1 = y_2 - y_3 = -1 \qquad b_2 = 1 \qquad b_3 = 0$$
$$c_1 = x_3 - x_2 = -1 \qquad c_2 = 0 \qquad c_3 = 1$$

$$k_{ij} = \frac{k}{4(1/2)} \begin{bmatrix} 2 & -1 & -1 \\ -1 & 1 & 0 \\ -1 & 0 & 1 \end{bmatrix} = \frac{k}{2} \begin{bmatrix} 2 & -1 & -1 \\ -1 & 1 & 0 \\ -1 & 0 & 1 \end{bmatrix} \qquad (5.46)$$

Quadrilateral Element

A quadrilateral (quad) element has four sides. Let us consider a quad element of rectangular shape having a length of $2l$ and a height of $2h$, as shown in Fig. 5.5.

The rectangular shape for the quadrilateral element is chosen to illustrate the formulation. Unlike the finite difference method, which must be orthogonal, there is no restriction of orthogonality of edges on the shape of the quad element. One must still follow the general guidelines on finite element meshes, such as aspect ratios, warp, and skewness, to get a good solution. The shape functions for the rectangular quad element are given by

$$\phi_1 = \frac{1}{4lh} (l - x)(h - y)$$

$$\phi_2 = \frac{1}{4lh} (l + x)(l - y)$$

$$\phi_3 = \frac{1}{4lh} (l - x)(h + y)$$

$$\phi_4 = \frac{1}{4lh} (l + x)(h + y) \qquad (5.47)$$

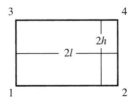

FIGURE 5.5. A rectangular quad element.

The physical region in most industrial thermal problems is not straight-sided and orthogonal, and therefore the shape functions are defined in terms of the elemental or natural coordinate system (χ, η). For the rectangular case

$$\chi = \frac{x}{l} \qquad \eta = \frac{y}{h}$$

χ and η are local coordinates, and they replace global coordinates. Therefore,

$$\phi_1 = \frac{1}{4}(1 - \chi)(1 - \eta)$$

$$\phi_2 = \frac{1}{4}(1 + \chi)(1 - \eta)$$

$$\phi_3 = \frac{1}{4}(1 - \chi)(1 + \eta) \qquad -1 \le \chi \le +1 \qquad -1 \le \eta \le +1$$

$$\phi_4 = \frac{1}{4}(1 + \chi)(1 + \eta) \tag{5.48}$$

If the coordinate system is located at the left corner of the element instead of at the center, the shape functions are given as

$$\phi_1 = (1 - \chi)(1 - \eta)$$
$$\phi_2 = \chi(1 - \eta)$$
$$\phi_3 = (1 - \chi)\eta \qquad 0 \le \chi \le \qquad 0 \le \eta \le 1$$
$$\phi_4 = \chi\eta \tag{5.49}$$

The conductivity matrix after the coordinate transformation from (x, y) to (χ, η) is

$$K = \int\int_{\text{Area}} k\left[\frac{\partial\phi_i}{\partial x}\frac{\partial\phi_j}{\partial x} + \frac{\partial\phi_i}{\partial y}\frac{\partial\phi_j}{\partial y}\right] dx\,dy$$

$$= \int_{-1}^{1}\int_{-1}^{1} k\left[\frac{\partial\phi_i}{\partial \chi}\frac{\partial\phi_j}{\partial \chi} + \frac{\partial\phi_i}{\partial \eta}\frac{\partial\phi_j}{\partial \eta}\right] |J|d\chi\,d\eta \tag{5.50}$$

where J is the Jacobian of coordinate transformation

$$J = \begin{bmatrix} \dfrac{\partial x}{\partial \chi} & \dfrac{\partial y}{\partial \chi} \\[2mm] \dfrac{\partial x}{\partial \eta} & \dfrac{\partial y}{\partial \eta} \end{bmatrix} \tag{5.51}$$

The mass matrix is given by

$$\iint_{\text{Area}} \phi_i\phi_j dx\, dy = \int_{-1}^{1}\int_{-1}^{1} \phi_i\phi_j |J| d\chi d\eta \qquad (5.52)$$

The integrals in (χ, η) coordinates are not straightforward, and usually the integrals are evaluated numerically with Gaussian quadratures. The Gaussian quadrature transforms the integral into a weighted sum over finite points known as Gauss points.

$$\int_{-1}^{1} f(\eta)d\eta \approx \sum_{j=1}^{n} w_j f(\eta_j) \qquad (5.53)$$

where w_j are weights, and η_j are the Gauss points.

$$\int_{-1}^{1}\int_{-1}^{1} f(\chi, \eta)d\chi\, d\eta = \sum_{i=1}^{n}\sum_{j=1}^{n} w_i w_j f(\chi_i, \eta_j) \qquad (5.54)$$

For the linear quad element $n = 2$ and weights $w_i = w_j = 1.0$, the Gauss points are

$$\chi_1 = \eta_1 = \frac{1}{\sqrt{3}} \qquad \chi_2 = \frac{1}{\sqrt{3}} \qquad \eta_2 = -\frac{1}{\sqrt{3}}$$

$$\chi_3 = -\frac{1}{\sqrt{3}} \qquad \eta_3 = \frac{1}{\sqrt{3}} \qquad \chi_4 = \eta_4 = \frac{1}{\sqrt{3}} \qquad (5.55)$$

Let us look at a simple rectangular quad element of length l and height h. The shape functions with the element coordinate system at the left-hand corner

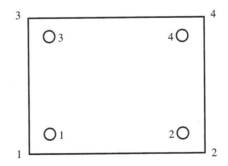

FIGURE 5.6. Quad element with Gauss points (integration points).

of the element are given by Eq. (5.49). The conductivity matrix for this element is evaluated from the following integral:

$$\left[\int_0^1 \int_0^1 \left(k\, \frac{\partial \phi_i}{\partial \chi}\frac{\partial \phi_j}{\partial \chi} + k\, \frac{\partial \phi_i}{\partial \eta}\frac{\partial \phi_j}{\partial \eta} \right) J|d\chi\, d\eta \right] \tag{5.56}$$

The Jacobian transformation is given by

$$J = \begin{bmatrix} h & 0 \\ 0 & l \end{bmatrix} \qquad J^{-1} = \frac{1}{lh}\begin{bmatrix} l & 0 \\ 0 & h \end{bmatrix}$$

Therefore,

$$\begin{bmatrix} \dfrac{\partial \phi_i}{\partial x} \\[2mm] \dfrac{\partial \phi_i}{\partial y} \end{bmatrix} = \frac{1}{lh}\begin{bmatrix} l & 0 \\ 0 & h \end{bmatrix}\begin{bmatrix} \dfrac{\partial \phi}{\partial \chi} \\[2mm] \dfrac{\partial \phi}{\partial \eta} \end{bmatrix} = \begin{bmatrix} \dfrac{1}{h}\dfrac{\partial \phi_i}{\partial \chi} \\[2mm] \dfrac{1}{l}\dfrac{\partial \phi_i}{\partial \eta} \end{bmatrix} \tag{5.57}$$

Differentiating the shape functions, we get

$$\frac{\partial \phi_1}{\partial \chi} = -(1-\eta) \qquad \frac{\partial \phi_2}{\partial \chi} = (1-\eta) \qquad \frac{\partial \phi_3}{\partial \chi} = -\eta \qquad \frac{\partial \phi_4}{\partial \chi} = \eta$$

$$\frac{\partial \phi_1}{\partial \eta} = -(1-\chi) \qquad \frac{\partial \phi_2}{\partial \eta} = -\chi \qquad \frac{\partial \phi_3}{\partial \eta} = (1-\chi) \qquad \frac{\partial \phi_4}{\partial \eta} = \chi$$

Therefore,

$$\int_0^1 \int_0^1 k\,\frac{\partial \phi_i}{\partial \chi}\frac{\partial \phi_j}{\partial \chi} |J|d\chi\, d\eta$$

$$= klh\,\frac{1}{h^2}\int_0^1\int_0^1 \begin{bmatrix} (1-\eta)^2 & -(1-\eta)^2 & \eta(1-\eta) & -\eta(1-\eta) \\ -(1-\eta)^2 & (1-\eta)^2 & -\eta(1-\eta) & \eta(1-\eta) \\ \eta(1-\eta) & -\eta(1-\eta) & \eta^2 & -\eta^2 \\ -\eta(1-\eta) & \eta(1-\eta) & -\eta^2 & \eta^2 \end{bmatrix} d\chi\, d\eta$$

$$= \frac{1}{6}\,k\,\frac{l}{h}\begin{bmatrix} 2 & -2 & 1 & -1 \\ & 2 & -1 & 1 \\ \text{symmetric} & & 2 & -2 \\ & & & 2 \end{bmatrix} \tag{5.58}$$

Similarly,

$$\int_0^1 \int_0^1 k \, \frac{\partial \phi_i}{\partial \eta} \, \frac{\partial \phi_j}{\partial \eta} \, |J| d\chi \, d\eta = \frac{1}{6} \, k \, \frac{l}{h} \begin{bmatrix} & & 2 & 1 & -2 & -1 \\ & & & 2 & -1 & -2 \\ & \text{symmetric} & & & 2 & 1 \\ & & & & & 2 \end{bmatrix} \qquad (5.59)$$

The combined conductivity matrix is

$$\frac{1}{6} \, k \, \frac{l}{h} \begin{bmatrix} & & 4 & -1 & -1 & -2 \\ & & & 4 & -2 & -1 \\ & \text{symmetric} & & & 4 & -1 \\ & & & & & 4 \end{bmatrix} \qquad (5.60)$$

THREE-DIMENSIONAL ELEMENTS

Tetrahedron Element

A linear three-dimensional tetrahedron element has four nodes. Similar to the formulation for the triangular element, where an area coordinate system is used, a volume coordinate system is used for the tetrahedron (tet) element, as follows:

$$L_i = V_i / V$$

V_i is the volume fraction at node i and V is the total volume of the element, so that

$$\sum_{i=1}^{4} L_i = 1.0$$

The linear shape functions for the tet element are

$$\phi_1 = L_1$$
$$\phi_2 = L_2$$
$$\phi_3 = L_3$$
$$\phi_4 = L_4 \qquad (5.61)$$

The mass matrix is once again defined as

$$M = \int_{\text{Volume}} \phi_i \phi_j \, dV$$

$$= \int_V \begin{bmatrix} L_1 \\ L_2 \\ L_3 \\ L_4 \end{bmatrix} [L_1 \quad L_2 \quad L_3 \quad L_4] dx \, dy \, dz$$

$$= \int_V \begin{bmatrix} L_1^2 & L_1L_2 & L_1L_3 & L_1L_2 \\ & L_2^2 & L_2L_3 & L_2L_4 \\ \text{symmetric} & & L_3^2 & L_3L_4 \\ & & & L_4^2 \end{bmatrix} dx \, dy \, dz \qquad (5.62)$$

The above volume integral can be evaluated from the following formula based on volume coordinates:

$$\int_V L_1^a L_2^b L_3^c L_4^d = \frac{a! \, b! \, c! \, d!}{(3 + a + b + c + d)!} \, 6V \qquad (5.63)$$

The integral in Eq. (5.62) therefore reduces to

$$\frac{V}{20} \begin{bmatrix} 2 & 1 & 1 & 1 \\ & 2 & 1 & 1 \\ \text{symmetric} & & 2 & 1 \\ & & & 2 \end{bmatrix} \qquad (5.64)$$

The conductivity matrix is given by

$$K = \int_V k \left[\frac{\partial \phi_i}{\partial x} \frac{\partial \phi_j}{\partial x} + \frac{\partial \phi_i}{\partial y} \frac{\partial \phi_j}{\partial y} + \frac{\partial \phi_i}{\partial z} \frac{\partial \phi_j}{\partial z} \right] dx \, dy \, dz$$

$$= \frac{k}{36V} \left\{ \begin{bmatrix} a_1^2 & a_1a_2 & a_1a_3 & a_1a_4 \\ & a_2^2 & a_2a_3 & a_2a_4 \\ \text{symmetric} & & a_3^2 & a_3a_4 \\ & & & a_4^2 \end{bmatrix} \right.$$

$$+ \begin{bmatrix} b_1^2 & b_1b_2 & b_1b_3 & b_1b_4 \\ & b_2^2 & b_2b_3 & b_2b_4 \\ \text{symmetric} & & b_3^2 & b_3b_4 \\ & & & b_4^2 \end{bmatrix}$$

$$+ \left. \begin{bmatrix} c_1^2 & c_1c_2 & c_1c_3 & c_1c_4 \\ & c_2^2 & c_2c_3 & c_2c_4 \\ \text{symmetric} & & c_3^2 & c_3c_4 \\ & & & c_4^2 \end{bmatrix} \right\}$$

where

$$a_1 = (y_2 - y_4)(z_3 - z_4) - (y_3 - y_4)(z_3 - z_4)$$
$$a_2 = (y_3 - y_4)(z_1 - z_4) - (y_1 - y_4)(z_3 - z_4)$$
$$a_3 = (y_1 - y_4)(z_2 - z_4) - (y_2 - y_4)(z_1 - z_4)$$
$$a_2 = -(a_1 + a_2 + a_3)$$

$$b_1 = (x_3 - x_4)(z_2 - z_4) - (x_2 - x_4)(z_3 - z_4)$$
$$b_2 = (x_1 - x_4)(z_3 - z_4) - (x_3 - x_4)(z_1 - z_4)$$
$$b_3 = (x_2 - x_4)(z_1 - z_4) - (x_1 - x_4)(z_2 - z_4)$$
$$b_4 = -(b_1 + b_2 + b_3)$$

$$c_1 = (x_2 - x_4)(y_3 - y_4) - (x_3 - x_4)(y_2 - y_4)$$
$$c_2 = (x_3 - x_4)(y_1 - y_4) - (x_1 - x_4)(y_3 - y_4)$$
$$c_3 = (x_1 - x_4)(y_2 - y_4) - (x_2 - x_4)(y_1 - y_4)$$
$$c_4 = -(c_1 + c_2 + c_3) \tag{5.65}$$

If there is heat generation inside the tet element, then the right-hand side is

$$F = \int_V \dot{q}\phi_i\, dx\ dy\ dz = \int_V \dot{q}\begin{bmatrix} \phi_1 \\ \phi_2 \\ \phi_3 \\ \phi_4 \end{bmatrix} dx\ dy\ dz$$

$$= \int_V \dot{q}\begin{bmatrix} L_1 \\ L_2 \\ L_3 \\ L_4 \end{bmatrix} dx\ dy\ dz = \frac{\dot{q}V}{4}\begin{bmatrix} 1 \\ 1 \\ 1 \\ 1 \end{bmatrix} \tag{5.66}$$

Hexahedron Element

A linear hexahedron (hex) element has eight nodes. The shape functions in the element coordinate system are given as

$$\phi_1 = \frac{1}{8}\ (1 - \chi)(1 - \eta)(1 - \tau)$$

$$\phi_2 = \frac{1}{8}\ (1 + \chi)(1 - \eta)(1 - \tau)$$

$$\phi_3 = \frac{1}{8}\,(1+\chi)(1+\eta)(1-\tau)$$

$$\phi_4 = \frac{1}{8}\,(1-\chi)(1+\eta)(1-\tau)$$

$$\phi_5 = \frac{1}{8}\,(1-\chi)(1-\eta)(1+\tau)$$

$$\phi_6 = \frac{1}{8}\,(1+\chi)(1-\eta)(1+\tau)$$

$$\phi_7 = \frac{1}{8}\,(1+\chi)(1+\eta)(1+\tau)$$

$$\phi_8 = \frac{1}{8}\,(1-\chi)(1+\eta)(1+\tau) \tag{5.67}$$

Using the Galerkin method described earlier, one can get the mass and conductivity matrices. The mass matrix is usually lumped, i.e., all the nondiagonal entries in a row are added, and then they are lumped on the diagonal. For the orthogonal hex element $1 \times 1 \times 1$, the entry on the diagonal is $1/8$. The conductivity matrix for the $1 \times 1 \times 1$ cube hex element shown in Fig. 5.7 is

$$\frac{k}{12}
\begin{bmatrix}
4 & 0 & 0 & -1 & 0 & -1 & -1 & -1 \\
& 4 & -1 & 0 & -1 & 0 & -1 & -1 \\
& & 4 & 0 & -1 & -1 & 0 & -1 \\
& & & 4 & -1 & -1 & -1 & 0 \\
& & & & 4 & 0 & 0 & -1 \\
\text{symmetric} & & & & & 4 & -1 & 0 \\
& & & & & & 4 & 0 \\
& & & & & & & 4
\end{bmatrix} \tag{5.68}$$

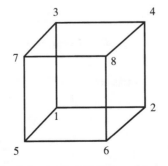

FIGURE 5.7. A typical orthogonal hex element.

Example 5.2

Formulate the conductivity and mass matrices for a tetrahedron element with the vertices of the element as shown in Fig. 5.8. Assume thermal conductivity k, density ρ and specific heat C_p to be constant:

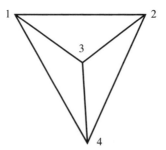

FIGURE 5.8. A tetrahedral element with node 1 being (0, 1, 0); node 2, (1, 1, 1); node 3, (0, 1, 1); and node 4, (0, 0, 1).

The conductivity matrix, given by Eq. (5.65), is

$$\frac{k}{36V}\begin{bmatrix} a_1^2 + b_1^2 + c_1^2 & a_2a_1 + b_2b_1 + c_2c_1 & a_3a_1 + b_3b_1 + c_3c_1 & a_1a_4 + b_1b_4 + c_1c_4 \\ & a_2^2 + b_2^2 + c_2^2 & a_3a_2 + b_3b_2 + c_3c_2 & a_2a_4 + b_2b_4 + c_2c_4 \\ \text{symmetric} & & a_3^2 + b_3^2 + c_3^2 & a_3a_4 + b_3b_4 + c_3c_4 \\ & & & a_4^2 + b_4^2 + c_4^2 \end{bmatrix}$$

The variables a, b, and c are given by Eq. (5.65) and, after substituting the nodal coordinates, are

$$a_1 = 0 \quad\quad a_2 = -1 \quad\quad a_3 = 1 \quad\quad a_4 = 0$$
$$b_1 = 0 \quad\quad b_2 = 0 \quad\quad b_3 = -1 \quad\quad b_4 = 1$$
$$c_1 = 1 \quad\quad c_2 = 0 \quad\quad c_3 = -1 \quad\quad c_4 = 0$$

$$V = \text{volume of element} = 0.166680$$

Therefore, the conductivity matrix for the tet element is

$$\frac{k}{36\,(0.166680)}\begin{bmatrix} 1 & 0 & -1 & 0 \\ & 1 & -1 & 0 \\ & & 3 & -1 \\ \text{symmetry} & & & 1 \end{bmatrix}$$

The mass matrix is

$$\frac{V}{20} \begin{bmatrix} 2 & 1 & 1 & 1 \\ & 2 & 1 & 1 \\ \text{symmetry} & & 2 & 1 \\ & & & 2 \end{bmatrix}$$

Lumping the mass matrix, we get

$$\frac{0.166680}{20} \begin{bmatrix} 5 & 0 & 0 & 0 \\ & 5 & 0 & 0 \\ \text{symmetric} & & 5 & 0 \\ & & & 5 \end{bmatrix}$$

Example 5.3

For the tet element in Example 5.2, find a steady state solution if node 2 is held at a constant temperature of 300°K and heat of 5 W is put in at node 4.

The system of equations becomes

$$\frac{1}{36\,(0.16668)} \begin{bmatrix} 1 & 0 & -1 & 0 \\ 0 & 1 & -1 & 0 \\ -1 & -1 & 3 & -1 \\ 0 & 0 & -1 & 1 \end{bmatrix} \begin{bmatrix} T_1 \\ T_2 \\ T_3 \\ T_4 \end{bmatrix} = \begin{bmatrix} 0 \\ 0 \\ 0 \\ 5 \end{bmatrix}$$

As temperature at node 2 is known, we can eliminate the second equation from the above system. A simple way to achieve this is to add a large number on the diagonal entry for node 2 and add the product of the large number and the temperature constraint on the right-hand side at row 2. Solving the system, we get

$$T_1 = 330°\text{K} \qquad T_2 = 300°\text{K} \qquad T_3 = 330°\text{K} \qquad T_4 = 360°\text{K}$$

This chapter briefly introduced the finite element method as applied to thermal problems. The concepts were presented in just enough detail to highlight the underlying principles. For a more detailed treatment of the subject, the readers are referred to [7] and [8].

The conductivity matrix and the mass matrix generated by the finite element method discussed in this chapter can be used to create conductors and capacitors for an equivalent thermal network. The thermal network thus constructed can be solved by iterative or direct solution methods. The methodology to create an equivalent thermal network from the finite element formulation is discussed in the next chapter.

Chapter 6
The Hybrid Method

This chapter describes a novel numerical method that combines the geometric flexibility of the finite element method (FEM) with the computational efficiency and the ease and clarity of applying complex boundary conditions of the finite difference or the control volume methods (FDM and CVM) described in Chapters 3 and 4.

Engineering analysis can be broken into three stages: input, compute, and output. During the input stage, an analyst prepares or describes the model for analysis. In the modern, computer-aided engineering (CAE) environment, this step is referred to as preprocessing, a time when the engineer/analyst describes the geometry of the physical domain, followed by mapping the geometry with a finite element mesh and then describing material properties, element properties, and boundary conditions. There are many commercially available graphical preprocessors, such as Patran™, Ideas™, etc., that have very powerful and easy-to-use geometry and finite element creation tools. Structural analysts have been using these finite-element-based preprocessors successfully in industry for the last two decades. In an integrated CAE environment, it is highly desirable that analysts of various disciplines be able to share and work off the same models. Therefore, a strong case can be made for using finite elements for thermal analysis. Finite element solution techniques are, however, not efficient, especially for nonlinear thermal problems. The following simple benchmark illustrates the inefficiencies of finite-element-based software.

The problem consists of an iron cube that is 1 m × 1 m × 1 m in dimension (Fig. 6.1). A constant temperature of 1500°K is imposed on the top face of the cube, and the bottom face is kept at 300°K. The problem is solved for constant as well as temperature-dependent thermal conductivity. The problem is run on a Silicongraphics workstation having a 5.3 operating system, 128 Mb of memory, and 700 Mb of disk space. Computer programs using a direct Finite Element Analysis (FEA) solution and a hybrid method with an iterative solver are used to run the problem. Pre- and post-processing are performed with Patran™. A CPU comparison for various mesh densities is performed, and the results are shown in Table 6.1.

One can observe from Table 6.1 that a direct/FEA solution is quite inefficient, especially for a nonlinear problem where the thermal conductivity of the iron is temperature-dependent. As the mesh was refined to be 41 × 41 × 41 and beyond, the finite-element-based programs ran out of disk space and could not solve the problem. The machine had about 700 megabytes of disk space available. The hybrid method using the iterative solution technique does not store the matrix, and it solved the problem quite efficiently, even for a 51 × 51 × 51 mesh density and did not run into disk space problems. The CPU times for the nonlinear prob-

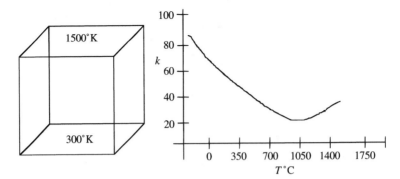

FIGURE 6.1. Boundary conditions and temperature-dependent thermal conductivity for the iron cube problem.

lem with the hybrid method using iterative solution are close to the ones for the linear problem.

To take advantage of the efficiency of finite-difference-type iterative solution techniques and finite-element-based preprocessors, one of the proposed methods was to use the finite element mesh and make nodes at the centroids of the elements and then connect them with thermal conductors and have capacitors attached to each of these centroidal nodes. The resistor-capacitor (R-C) network thus created was fed into an R-C network solver. The limitations and problems associated with the element centroidal approach are discussed below.

Figure 6.2 shows two elements, 1 and 2, with centroidal nodes 1 and 2. The heat flows from node 1 to 2 and can be written as

$$q_{1-2} = \frac{k_{1-2}A(T_1 - T_2)}{L}$$

Table 6.1. CPU Time Comparison for the Iron Cube Problem

Mesh Density	Constant Conductivity		Variable Conductivity	
	Hybrid/ Iterative	Direct/FEA	Hybrid/ Iterative	Direct/FEA
11 × 11 × 11	6.91 sec	34.45 sec	8.34 sec	34.47 sec
21 × 21 × 21	1 min 20 sec	3 min 38 sec	1 min 41 sec	4 min 7 sec
31 × 31 × 31	5 min 44 sec	57 min 52 sec	8 min 3 sec	57 min 35 sec
		Out of disk space		Out of disk space
41 × 41 × 41	21 min 1 sec		27 min 32 sec	
		Out of disk space		Out of disk space
51 × 51 × 51	1 hr 6 min 9 sec		1 hr 24 min	

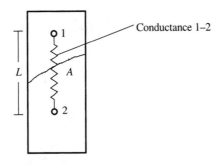

FIGURE 6.2. A typical conductance for an element centroidal technique.

If elements 1 and 2 are made of dissimilar materials, the thermal conductivity k_{1-2} has to be approximated as some combination of the thermal conductivity k of both material 1 and 2. This can lead to inaccuracies if the thermal conductivities of the two materials are orders of magnitude apart (e.g., conductor and insulator) or when the conductivities are strongly dependent on temperature. Creating nodes at the element centroids also increases bookkeeping, as translation from these nodes to true finite element nodes has to be performed for the results to be evaluated.

A serious problem with the element centroid technique is that it can generate physically unrealistic results. In most industrial applications, the physical domain does not conform to the orthogonal heat flow paths inherent to the finite difference formulations. As one begins to skew the element, significant errors in solution are introduced in the finite difference element centroidal formulation. The finite element method uses a bilinear functional dependence within an element and therefore can describe correctly the temperature distribution even in highly skewed elements. The Kershaw problem illustrated below was designed to highlight the effect of skewed meshes. A rectangular plate is kept at a constant temperature of 0 on the bottom edge and a constant heat flux is applied at the top edge. A skewed finite element mesh is used to discretize the domain. Of course, one would not model a rectangular domain in this manner, but one can envision how one could generate skewed meshes modeling highly complex geometries. The results, shown in Fig. 6.3, are obtained by finite difference element centroidal technique [10]. It can be observed that the isotherms follow the mesh lines. Results obtained by finite-element-based software show physically realistic results [10].

One can conclude from the above discussion that finite elements have geometric flexibility, plus an ability to use true finite element nodes generated by the preprocessors with the solutions being physically realistic even for skewed meshes. However, finite element solutions suffer from computational inefficiency. This is the motivation for the novel hybrid method that combines the advantages of finite element modeling with the computational efficiency of finite difference or resistor-capacitor network solvers. This method is implemented in the commer-

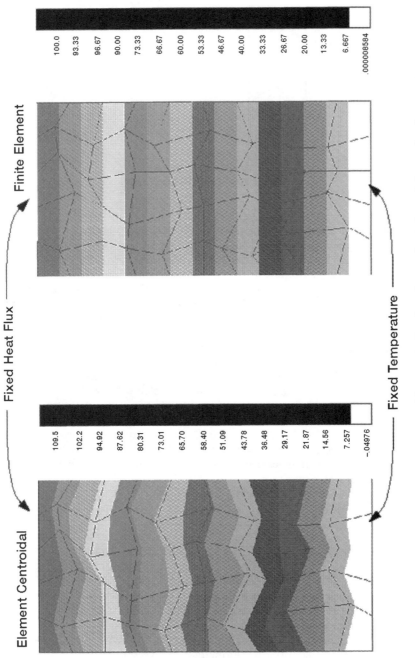

FIGURE 6.3. Kershaw problem, results for skewed meshes: (a) results for element centroidal technique; (b) results from a finite-element-based code.

cially available thermal analysis package MSC/Patran/Thermal^{TM}, which was previously known as P/Thermal.

FINITE ELEMENT TO RESISTOR-CAPACITOR METHOD

Figure 6.4 shows the basic philosophy of the hybrid method. We want to convert the finite elements into an equivalent resistor–capacitor or conductance–capacitor network. At each node, we will have various conductances coming from each of the adjoining nodes and a capacitance, with each element connected to the node contributing a part of that capacitance. If the elements connected to the node are made of same material, then the capacitances are merged; otherwise, they are kept separate. Similarly, conductances are merged only if they have the same material associated with them. Therefore, no approximation needs to be made for interfacial conductivity as long as the model is constructed so that no element crosses over another material. This is a trivial modeling requirement for the robust preprocessors available commercially.

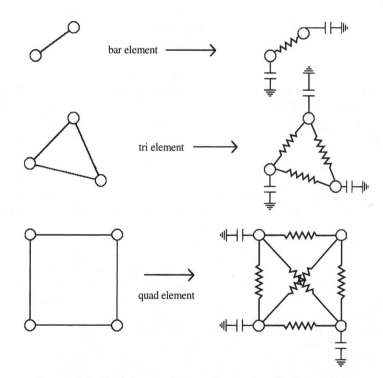

FIGURE 6.4. Resistor and capacitor network for various finite elements.

Let us look at a triangular element. Thermal conductivity matrix for this element as derived in the previous chapter for isotropic thermal conductivity is

$$[K] = \frac{k}{4A^e} \begin{bmatrix} a_1^2 + b_1^2 & b_1 b_2 + a_1 a_2 & a_1 a_3 + b_1 b_3 \\ a_1 a_2 + b_1 b_2 & b_2^2 + a_2^2 & a_2 a_3 + b_2 b_3 \\ a_1 a_3 + b_1 b_3 & a_2 a_3 + b_2 b_3 & a_3^2 + b_3^2 \end{bmatrix}$$

where a and b are

i	a_i	b_i
1	$y_2 - y_3$	$x_2 - x_3$
2	$y_3 - y_1$	$x_1 - x_3$
3	$y_1 - y_2$	$x_2 - x_1$

(6.1)

Consider a conductance network between the same set of nodes. When assembled in a matrix form, $[K]\{T\} = \{b\}$ and so

$$[K] = \begin{bmatrix} G_{12} + G_{13} & -G_{12} & -G_{13} \\ -G_{12} & G_{12} + G_{23} & -G_{23} \\ -G_{13} & -G_{23} & G_{13} + G_{23} \end{bmatrix}$$

(6.2)

This conductivity matrix is similar to the one derived from the finite element theory. The matrix is symmetric, and the sum of off-diagonal terms is the negative of the diagonal term. The conductor network described in Eq. (6.2) is equivalent to the conductivity matrix of the triangular element from the finite element formulation if

$$k_{ij} = -G_{ij} \qquad i \neq j$$

(6.3)

This equation is not restricted to a triangular element, but can be applied to any element [23]. Therefore, any finite element thermal model can be converted into an equivalent conductance-capacitor network. The capacitance is computed using the same methods used in finite elements to calculate the lumped capacitance matrix. Once the conductance-capacitance network is formulated, one can solve the network using efficient network solution methodologies, unlike the finite element codes, which solve the linear system of equations $[K]\{T\} = \{F\}$. The finite element using direct solution requires assembly of element matrices into a global conductivity matrix, then factoring the matrix and solving using (usually) some kind of Gaussian elimination technique or other matrix solution techniques. For nonlinear problems with nonlinear thermal conductivity, the element matrices usually have to be recomputed for each iteration and reassembled and factored and solved. When the problem is highly nonlinear and also transient, the process has to be repeated for every time interval. That is why the programs using direct solution techniques tend to be computationally inefficient for highly nonlinear problems. As the problem size increases—that is, the number

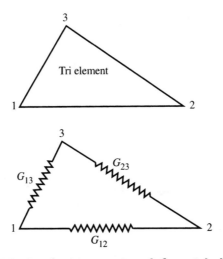

FIGURE 6.5. Conductance network for a tri element.

of nodes gets large—the matrix solution cannot be performed in available computer memory (in core), and the matrix is stored on disk and solved out of core. This increases the CPU time required for solution. For moderately sized problems (5000–7000 nodes), the direct solution technique used in finite element programs is on par with iterative network solvers, at least for mildly nonlinear problems. In some so-called "stiff" thermal problems, direct solutions may be even faster than the iterative network solution techniques. A comparative analysis of direct versus iterative solution techniques is discussed later in this chapter. Let us now study a couple of example problems to illustrate how the hybrid FE/FD technique can be used to solve thermal problems.

Example 6.1
Let us study a square plate 1 m × 1 m, with constant thermal conductivity k = 1 W/m · K. Use one quadrilateral element. Heat of 1 W is input at both the left corner and the top right corner, and the temperature is held constant at 25°C (Fig. 6.6).

The conductance network is generated from the finite element matrix for this quadrilateral element. From the previous chapter, the element matrix for this quad element is given by Eq. (5.60) as

$$\frac{1}{6} \begin{bmatrix} 4 & -1 & -2 & -1 \\ & 4 & -1 & -2 \\ & & 4 & -1 \\ \text{symmetric} & & & 4 \end{bmatrix} \begin{matrix} 1 \\ \\ \\ \end{matrix}$$

Therefore, from Eq. (6.3) the conductances (G) are given by

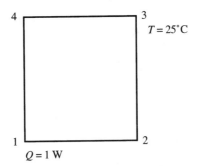

FIGURE 6.6. Quadrilateral element with node numbers and boundary conditions.

$$G_{1-2} = -\left(-\frac{1}{6}\right) = 0.16666$$

$$G_{1-4} = -\left(-\frac{1}{6}\right) = 0.16666$$

$$G_{1-3} = -\left(-\frac{2}{6}\right) = 0.33333$$

$$G_{2-3} = -\left(-\frac{1}{6}\right) = 0.16666$$

$$G_{2-4} = -\left(-\frac{2}{6}\right) = 0.33333$$

$$G_{3-4} = -\left(-\frac{1}{6}\right) = 0.16666$$

These conductances are assembled in a matrix form, and we apply the nodal heat and temperature boundary conditions, as follows:

$$\begin{bmatrix} 0.6666 & -0.1666 & -0.33333 & -0.1666 \\ -0.1666 & 0.6666 & -0.1666 & -0.3333 \\ 0.3333 & -0.1666 & 0.6666+n & -0.1666 \\ 0.1666 & 0.3333 & -0.1666 & 0.6666 \end{bmatrix} \begin{bmatrix} T_1 \\ T_2 \\ T_3 \\ T_4 \end{bmatrix} = \begin{bmatrix} 1 \\ 0 \\ 25^*n \\ 0 \end{bmatrix}$$

The temperature at node 3 is known ($T_3 = 25°C$). An easy way to introduce a temperature constraint without eliminating the equation for that node is to add a large number n (say $n = 10^8$) on the diagonal and multiply the temperature constraint by the same large number (n) and then add it to the right-hand side.

The solution of the above matrix system is

$$T_1 = 27°C \qquad T_2 = 26°C \qquad T_3 = 25°C \qquad T_4 = 26°C$$

We can also use the iterative technique to solve the conductance network. The iterative method does not store any matrix, but it works on a node-by-node basis. Only one node is iterated at a time. One node is allowed to float, and all the other nodal temperatures are assumed to be known from a guessed value or previous iteration. For transient problems, one starts with the initial conditions and no guess about starting temperatures is required. For a steady state problem, we try to find, in essence, the temperature that will zero out the net heat flows from surrounding nodes into the node being iterated. Let us apply this procedure to Example 6.1. Let the iterations begin with a guessed temperature at 25°C for all the nodes.

Node 1:

$$(T_1 - T_3) * \underbrace{0.3333 * k}_{\text{conductance}} + (T_1 - T_2) * 0.1666 * k$$

$$+ (T_1 - T_4) * 0.1666 * k - \underbrace{1}_{\text{nodal heat}} = 0$$

$$(T_1 - 25) * \underbrace{0.3333 * 1}_{\text{conductance}} + (T_1 - 25) * 0.1666 * 1$$

$$+ (T_1 - 25) * 0.1666 * 1 - \underbrace{1}_{\text{nodal heat}} = 0$$

Therefore,

$$T_1 = 26.498$$

We use the new value for T_1 immediately for subsequent calculations.

Node 2:

$$(T_2 - 26.498) * 0.1666 + (T_4 - 25) * 0.1666 + (T_4 - 25) * 0.3333 = 0$$
$$T_2 = 25.374$$

Node 4:

$$(T_4 - 26.498) * 0.1666 + (T_4 - 25) * 0.1666 + (T_4 - 25.374) * 0.3333 = 0$$
$$T_4 = 25.56$$

We can repeat the process until the values do not change within a tolerance and obtain the temperature solution as

$$T_1 = 26.999 \qquad T_2 = 25.999 \qquad T_3 = 25 \qquad T_4 = 25.999$$

Let us now solve the above example problem using the two triangular elements shown. The conductivity matrix for the right triangle element with unit base and height is

$$\begin{bmatrix} 1 & -0.5 & -0.5 \\ & 0.5 & 0 \\ \text{symmetric} & & 0.5 \end{bmatrix}$$

Therefore, the conductance (G) are

$$G_{1-2} = 0.5$$
$$G_{2-4} = 0.5$$
$$G_{1-4} = 0$$
$$G_{1-3} = 0.5$$
$$G_{3-4} = 0.5$$

Assembling the conductances in a matrix form, we get

$$\begin{bmatrix} 1.0 & -0.5 & -0.5 & 0 \\ -0.5 & 1 & 0 & -0.5 \\ -0.5 & 0 & 1.0 & -0.5 \\ 0 & -0.5 & -0.5 & 1+n \end{bmatrix} \begin{bmatrix} T_1 \\ T_2 \\ T_3 \\ T_4 \end{bmatrix} = \begin{Bmatrix} 1 \\ 0 \\ 0 \\ 25*n \end{Bmatrix}$$

Again, as T_4 is known, a large number is added on the K (4, 4) term, and the known temperature is multiplied by the same large number and is added on the right-hand side. The solution of the matrix system gives the nodal temperatures

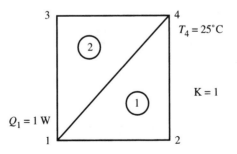

FIGURE 6.7. Triangular mesh for example 6.1 with boundary conditions.

$$T_1 = 26.999 \qquad T_2 = 25.999 \qquad T_3 = 25.999 \qquad T_4 = 25$$

which is the same solution as the one obtained with the quad element. (Note: Nodes 3 and 4 are switched for the quad and tri element problems, as explained.) The conductance or resistor network can be solved with a point iterative technique, as illustrated earlier. It is interesting to note that although the conductance network for the tri and quad elements is quite different, the two solutions are identical. The conductances generated by the hybrid method are purely mathematical; however, they model the physics correctly.

Let us now look at an example with a convective boundary condition.

Example 6.2
A 1 m × 1 m square plate is imposed with a heat flux (Q) of 1 W/m^2 on the left edge, and the right edge is convecting heat to the adjacent area (which has an ambient temperature of 25°C) with a heat transfer coefficient (h) of 5 W/m$^2 \cdot$ K. Mesh the square plate with one quad element and obtain the steady state temperature solution.

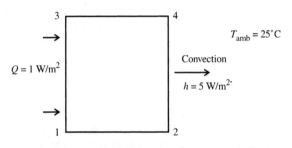

FIGURE 6.8. Quadrilateral element, nodes and boundary conditions for example 6.2.

The conductance network is as shown below:

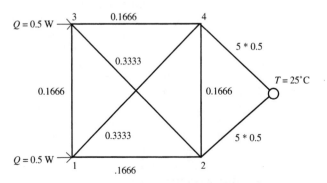

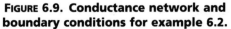

FIGURE 6.9. Conductance network and boundary conditions for example 6.2.

One can assemble the conductance network in matrix form and solve for temperatures, or one can solve the network with iterative techniques. The temperature solution is $T_1 = 26.199$, $T_2 = 25.199$, $T_3 = 26.199$, $T_4 = 25.199$.

The advantage of modeling convection with a fluid node and a convective conductance is that it gives flexibility in applying variable boundary conditions such as time-dependent temperature or a time/temperature-dependent convection coefficient. The convective conductance can be separated into the geometric part, the area, and the convection coefficient. Similarly, the conduction conductance, kA/L, can be separated into a geometric part, A/L, and material part, k. In the preceding examples, the thermal conductivity K was assumed constant; however, if the thermal conductivity is temperature-dependent, one can evaluate it at the average temperature of the two nodes constituting the conductance and multiply it by the geometric part to get the conductance. The geometric part has to be evaluated only once based on the finite element mesh. If direct solution techniques are employed, the temperature-dependent thermal conductivity has to be calculated every iteration and the matrix has to be assembled every iteration and solved. When using an iterative technique, no matrix assembly is required, and the temperature-dependent thermal conductivity is evaluated and immediately used while iterating on a node, and then one goes on to find the solution for the next node.

ITERATIVE AND DIRECT SOLUTION METHODS

The workings of an iterative method can be illustrated by Rockenbach's string computer [10, 11], as shown in Fig. 6.10. The string computer consists of a string pinned at one end and wrapped over a set of pegs placed at node locations. Attached to the other end of the string is a weight that keeps the string taught.

In Fig. 6.10(a), the pegs are uniformly spaced. The initial position of the string over the pegs can be thought of as an initial guess for the iterative solution, while the two pinned ends of the string represent boundary conditions. The iterations are started by removing peg 1; the string then repositions itself, and peg 1 is nailed at the new location of the string at the abscissa of the peg. Similarly, peg 2 is removed next and repositioned. This procedure is carried out from left to right and then from right to left until the string is flat and removing pegs no longer changes the position of the string. This is when one declares convergence of the iterations.

In Fig. 6.10(b), the pegs are not uniformly spaced; pegs 3 and 4 from the left are spaced very close to each other. It can be observed from Fig. 6.10(b) that the iterations march the string down very slowly to convergence. This is typical of a "stiff" thermal problem, for example, closely spaced nodes of a highly conducting material, or an insulator between two conductors, or a conductor between two insulators. In thermal analysis, such "stiff" problems can be very computationally intensive for iterative solvers, and direct solvers that assemble the matrix may work more efficiently. The iterative solvers work on a node-by-node basis and no matrix assembly of the equations is required. The direct solvers, on the other hand, assemble the conductance-capacitance network in a matrix form. MSC/Patran/Thermal^{TM}, for example, has both iterative and direct

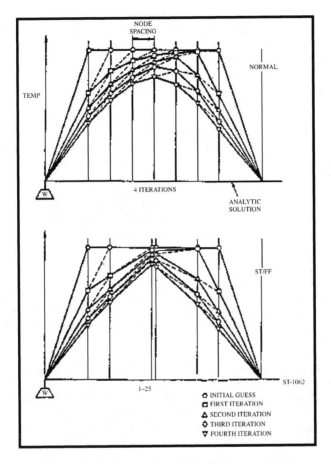

FIGURE 6.10. Rockenbach's string computer (a) equidistant pegs; (b) closely spaced pegs.

solvers. The direct solver in MSC/Patran/Thermal^{TM} uses the Choleski method, which is based on linear Gaussian elimination. The solver can solve the equations efficiently in core (all matrix coefficients fit in the computer memory) and out of core. The solver compresses out the zeros that are outside the skyline of the matrix, while the iterative solver is the Strongly Nonlinear Point Successive Over-Relaxation (SNPSOR) algorithm and is discussed later in this chapter. Three test cases are presented here to shed more light on the workings of direct and iterative solution methods. The cases were run with MSC/Patran/Thermal^{TM} on a VAX 8600 computer in 1991.

Test Case 1
The first test case is the same problem we used earlier to compare the relative performance of the direct/FEA solution versus the hybrid/iterative solution. The

problem consists of an iron cube 1 m × 1 m × 1 m, with constant temperature at 1500°K on the top face and a fixed temperature of 300°K on the bottom face; all other faces are adiabatic. The problem was run with a constant thermal conductivity of 30 W/m · K and a temperature-dependent thermal conductivity. Mesh sizes were of 2738, 3375, 6859, and 9261 nodes. The run time comparisons on a Vax 780 machine are tabulated in Table 6.2. Convergence criteria of 1. E −4 are used, and direct and iterative solvers of MSC Patran/Thermal^{TM} are used.

From the table, one can observe that direct solvers, which assemble the equations in a matrix form and solve them as a linear system of equations, tend to perform slowly for nonlinear problems and their performance deteriorates with increasing problem size. In general, when problem size is large (10,000+ nodes) and the nonlinearities of the problem require many iterations of computation (assembly, factoring, solving), the performance of direct solvers is inefficient. In such a situation, an iterative solver such as SNPSOR is highly recommended. On the other hand, for "stiff" thermal problems, the convergence of iterative methods is slow and direct solution might work better.

Test Case 2
The geometry and boundary conditions for test problem 2 is shown in Fig. 6.11. It consists of three 10 × 1 slabs and one 10 × 0.5 slab. The problem is meshed with 900 quadrilateral elements (1014 nodes). The problem is run with constant thermal conductivities. The problem is first run (case 2a) with all slabs at thermal conductivity $k = 1$ W/m · K. The direct solver took 37.29 seconds of CPU on a Vax 8600 computer, while the iterative solution consumed a CPU of 1 hour, 29 minutes, and 24.05 seconds. As the thermal conductivity is constant, only one iteration is required through the direct solver.

The iterations are started with an initial guess of 300°K, the constant temperature boundary condition. The iterative solver performs slowly, since the heat input has a long path to travel and it takes a lot of iterations (4505) to bring the information from the heat flux boundary to the interior of the solution domain.

Table 6.2. CPU Times for Test Case 1: Direct and Iterative Solvers

Number of Nodes (# resistors)	Variable Conductivity		Constant Conductivity	
	Direct Solver	Iterative Solver	Direct Solver	Iterative Solver
2738 (30 422)	46 min 42.81 sec	10 min 34.17 sec	5 min 11.22 sec	6 min 49.26 sec
3375 (38 066)	1 hr 20 min 15 sec	13 min 48.4 sec	8 min 37.96 sec	9 min 49.64 sec
6859 (79 758)	9 hr 20 min 42 sec	37 min 59.6 sec	51 min 3.36 sec	26 min 4.5 sec
9261 (108 860)	17 hr 16 min 16 sec	48 min 1.71 sec	1 hr 39 min 20 sec	37 min 12.6 sec

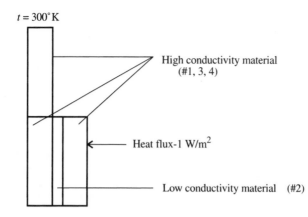

FIGURE 6.11. Geometry and boundary conditions for test case 2.

In case 2*b*, the thermal conductivity of the middle slab (slab #2) was chosen as 0.01. This situation is similar to a highly conducting material next to highly insulating material. As the thermal properties are constant, thermally the nodes in the conducting material are more closely spaced than the ones in the insulator, even though they are spaced equidistant physically. This situation gives rise to a thermally "stiff" problem, as was described earlier with the Rockenbach's string computer. Case 2*b* took 2 hours and 58 minutes, for the iterative solver which required 8843 iterations, while the direct solver took 37.2 seconds.

For case 2*c*, the problem becomes even more stiff because $k_1 = 1$, $k_2 = 0.01$, $k_3 = 10$, $k_4 = 1$. For this case, with 22,897 iterations, the iterative solver took 6 hours, 52 minutes, and 43.18 seconds, while the direct solver, needing only one iteration since the thermal conductivities are constant, took 37.5 seconds.

In case 2*d*, the top slab was removed, reducing the path from the flux boundary to constant temperature boundary. This case was run with 900 quad elements and 1014 nodes. The iterative solver required 961 iterations and took 11 minutes 52.26 seconds for a converged solution, while the direct solver ran in 27.08 seconds. Some researchers have proposed a block correction scheme for the iterative solvers for "stiff" problems. Such methods have been described for structured grids, and their performance on irregular meshes and complex geometries needs further investigation. A direct solution approach for such problems is recommended.

Test Case 3
As seen from cases 1 and 2, the direct solver performed poorly when material properties were temperature-dependent (case 1), while it outperformed the iterative solver for a thermally "stiff" problem (case 2) with constant material properties. Test case 3 consists of a copper cylinder 0.2 meter in diameter and 0.4 meter in length placed at the center of $1 \times 1 \times 1$ meter cube. The copper cylinder is generating 1000 W/m^3 and is radiating to the box kept at 300°K. Temperature-

dependent thermal conductivity of copper is used. The box is meshed with a 3×3 grid on all the faces, while the copper cylinder is meshed with 128 hex elements. The total number of nodes is 233. This is a stiff problem, as the resistance to flow of heat by conduction in copper is much less than the resistance to flow of heat by radiation from the copper cylinder to the enclosing box. The viewfactors and radiation resistors were calculated by the built-in viewfactor program within the MSC/Patran/Thermal™ system. When the problem was run with an iterative solver, the CPU time was 1 hour, 4 minutes, and 53.95 seconds (1805 iterations), and the direct solver ran it in 1 minute and 26.45 seconds (9 iterations of direct solution). All the computations were performed on a Vax 8600 computer. Although this problem has temperature-dependent properties and radiation, making the problem nonlinear, the direct solver outperformed the iterative solver. This is because the problem size was small (only 233 nodes) and the stiffness of the problem made it converge very slowly through the iterative solver.

TRANSIENT PROBLEMS

For transient problems, one can use a predictor-corrector algorithm. In the predictor part of the algorithm, linear extrapolation is used in the time domain to predict the future values of nodal temperatures. A simple way to extrapolate is by the difference in value of temperature at the current iteration level and the temperature at the previous time. The algorithm developed by Hughes [12], which is unconditionally stable for nonlinear problems, is used for the corrector step. Transient conduction involves the added complication of capacitors at node points. A capacitor is defined as

$$\text{Capacitor} = \text{Density} \cdot \text{Specific Heat} \cdot \text{Volume}$$
$$C = \rho C_p V$$

As we are using true element nodes, if the node in question lies at a material interface, one capacitor is assigned to the node for each material at the interface. The volume at a node is composed of contributions from all elements connected to the node. The volume is calculated from finite element shape functions (the same procedure as was used in the previous chapter). The capacitors at a node are merged only if they have the same material associated with them; otherwise they are kept separate.

The heat conduction equation solved for the transient problem at a given node i can be stated as

$$\sum_{i=1}^{\text{Nodes}} C[i] \frac{dT[i]}{dt} = \sum_{j=1}^{\text{Nodes}} \frac{T_j - T_i}{R_{ij}} \tag{6.4}$$

where $C[i]$ = Capacitance at node i
T = Temperature
t = Time
R_{ij} = Thermal resistance between nodes i and j

This states that the sum of all the capacitances multiplied by the time derivative of temperature at node i equals all the heat flows across all the resistors or conductors coming into node i. Equation (6.4) can be rearranged as

$$\frac{dT[i]}{dt} = \frac{\sum\limits_{i=1}^{\text{all}} Q[i]}{\sum\limits_{i=1}^{\text{all}} C[i]} \tag{6.5}$$

where $Q[i]$ = Heat flow into node i
C = Capacitance at node i

Therefore,

$$\frac{T[i,t+dt] - T[i,t]}{dt} = \frac{\sum\limits_{i=1}^{\text{all}} Q[i]}{\sum\limits_{i=1}^{\text{all}} C[i]} \tag{6.6}$$

One can evaluate the heat flows and capacitances at either the previous time level, t_{old} (explicit), the new time level, $t_{\text{old}} + dt$ (fully implicit), or somewhere in between. Therefore,

$$\frac{T(i,t+dt) - T(i,t)}{dt} = \frac{\sum\limits_{i=1}^{\text{all}} Q[i, T(t + \beta dt)]}{\sum\limits_{i=1}^{\text{all}} C[i, T(t + \beta dt)]} \tag{6.7}$$

where β is the Hughes integration algorithm explicit/implicit weighting factor and

$$T(i, t + \beta dt) = T[i,t](1 - \beta) + t[i, t + dt]\beta \tag{6.8}$$

If one treats heat absorbed into a capacitor as just another heat flow, the left-hand side of Eq. (6.7) can be moved to the right-hand side, and we get

$$\sum_{i=1}^{all} Q[i, T(t + \beta dt)] = 0 \tag{6.9}$$

The capacitor heat flow is given as $\rho C_p V \cdot dT$, where dT is the difference between the current iteration temperature and the temperature at the old time level. The problem therefore reduces to finding, for the node i that one is iterating, the temperature that will zero the net heat flow into that node. To compute the heat flows, the temperatures of the nodes are taken as the projected temperatures by the predictor

$$T_{node} = T_{old} \cdot (1 - \beta) + \beta T_{predicted} \tag{6.10}$$

Value of β is based on the explicit stable time interval for the node [13], as follows:

$$dt_{stable} = \frac{\sum \text{Capacitance}}{\sum \text{Conductance}} \text{ at a node} \tag{6.11}$$

$$R = \text{Current time interval/Explicit stable time step} \tag{6.12}$$

for $R \le 0.35$, $\beta = 0$; in other words, the problem is run as fully explicit otherwise

$$\beta = \frac{3R - 1}{6R} \quad \text{and} \quad \beta = \text{Max}(\beta, 0.0)$$

SUCCESSIVE OVERRELAXATION ALGORITHM (SOR)

We have seen in the previous section, on steady state thermal network, the use of a point iterative technique such as Gauss-Siedel to find the zero for the heat flow versus temperature curve. In Chapter 2, the Gauss-Siedel algorithm with overrelaxation was discussed. Such algorithms are adequate for mildly nonlinear problems. However, this algorithm can run into difficulties if the equation system is strongly nonlinear, such as is encountered with high-temperature radiation. When linear SOR algorithms encounter strongly nonlinear behavior, underrelaxation is used to force convergence [13]. However, underrelaxation usually causes convergence to be slow.

One of the approaches to this problem is to use a hybrid of Newton's first-order method and Newton's second-order method to estimate the zero of the Q versus T curve. Newton's first-order method is similar to the Gauss-Siedel method. Newton's second-order method uses three-function evaluations—$Q(T)$, $Q(T+\Delta T)$, $Q(T - \Delta T)$—to find the root of the Q versus T curve. ΔT is a small perturbation in temperature. Because Newton's second-order method requires three-function evaluations, it can be costly for large problems. Usually, conduction is mildly non-linear and conductive resistors are most numerous in large thermal problems. Therefore, the hybrid approach is used. Conduction is treated with first-order Newton's method (Gauss-Siedel). All other heat flows—e.g., convective, radiative, capacitances, heat sources—are treated with Newton's second-order method, i.e., three-function evaluations are calculated to estimate the zero of the Q versus T curve. Conductive conductances, which make up the majority of all conductances in a thermal model, are computed only once per iteration, while other heat flows are computed three times per iteration. This results in a fast and convergent algorithm capable of efficient computation for conduction problems and reliable convergence for strongly nonlinear problems. The SNPSOR algorithm discussed here does not, in general, require underrelaxation to achieve convergence. More information on Newton's second order method can be found in [13].

One can speed up the iteration process by using overrelaxation. The simplistic approach is to apply a fixed overrelaxation parameter between 1.0 and less than 2.0 to the entire system of equations. There is an optimal value of relaxation parameter that will result in the most efficient solution of the system of equations. Sometimes, an individual relaxation parameter applied to individual nodes may be the best approach. It is not easy to determine what overrelaxation parameter will result in the most efficient solution. It has been observed that estimating and updating the relaxation parameter as the iteration progresses leads to faster convergence than a fixed value of the relaxation parameter for the entire iteration process.

The overrelaxation parameter estimation algorithm described below is used in MSC/Patran/Thermal^{TM} and is heuristic in nature. It has evolved over the years of use of the program in various industrial thermal applications. The overrelaxation estimation algorithm is based on the iteration convergence rate R

$$R = \left| \frac{E(i)}{E(i-1)} \right| \tag{6.13}$$

where $E(i-1)$ is the largest iterative delta (positive or negative) at the $(i-1)$ iteration and $E(i)$ is the largest iterative delta at the i iteration. Iterative delta is defined as the change in nodal temperature for that iteration. In the limit as i goes to infinity, the following formula (see Eqs. 3-83, 3-113, 3-116 in [14]) is used to estimate the relaxation parameter.

$$A = \frac{(R + \text{Relax}_{current} - 1)}{\text{Relax}_{current}}$$

$$B = \frac{A^2}{R} \tag{6.14}$$

For $B \geq 1$, $\text{Relax}_{new} = \text{Relax}_{current}$. Otherwise,

$$C = \sqrt{1 - B}$$

$$\text{Relax}_{\text{new}} = \frac{2}{1 + C} \tag{6.15}$$

This relaxation parameter estimation algorithm works well as long as the iteration count approaches infinity and the eigenvalues of the iteration matrix have not become complex. In practice, the iteration count never approaches infinity (nor do we want it to) and eigenvalues can become complex. We are interested only in a good estimate for the convergence rate. The convergence rate will change drastically in early iterations. Therefore, the algorithm described above is not brought into play until the convergence rate has stabilized to an asymptotic limit. This can be done by requiring that for a finite number of iterations, say 4 or 5,

1) R is less than 1
2) The sign of iterative deltas has not changed
3) The difference in convergence rate is less than 1%.

If these conditions are met, we can assume that the convergence rate has stabilized.

The algorithm described above is only valid as long as eigenvalues of iteration matrix are not complex. How does one know when the eigenvalues of iteration matrix have become complex? Furthermore, as there is no theory for predicting the optimal relaxation parameter for a system with complex eigenvalues, how does one estimate a new relaxation parameter?

One indication of complex eigenvalues is that the largest iterative deltas $E(i)$ start oscillating in sign (positive and negative). Although the iterative deltas do change sign when eigenvalues are not complex, the oscillations in sign are continuous and frequent when eigenvalues are complex. Usually, the cause of complex eigenvalues is that the current relaxation value is greater than the optimum relaxation parameter. It is difficult to know how much greater. One remedy that can be used is to subtract a small value, say 0.001 or 0.002, from the current relaxation parameter every time the iterative deltas change sign. Sometimes, using a fixed relaxation parameter that is less than the relaxation parameter at which oscillations start can stabilize the iterations.

SKEWED MESHES AND NEGATIVE CONDUCTANCES

As mentioned earlier in the chapter, the hybrid FE/FD method solutions are physically realistic even for highly skewed meshes. This was illustrated with the benchmark Kershaw problem. For a 1 × 1 quad element, the geometric part of the conductances (A/L) is as follows:

$$G_{12} = G_{13} = G_{24} = G_{34} = 0.16666$$
$$G_{14} = G_{23} = 0.33333$$

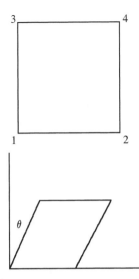

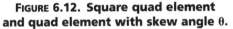

FIGURE 6.12. Square quad element and quad element with skew angle θ.

Let us start skewing the quad element, shown in Fig. 6.12. The angle of skew θ is increased 10° to 80°.

The conductances are tabulated below.

For θ = 10°:

$$G_{12} = 0.1692328, G_{23} = 0.4266353, G_{14} = 0.2503152, G_{13} = 0.1692420$$
$$G_{24} = 0.1692420, G_{34} = 0.1692429$$

For θ = 20°:

$$G_{12} = 0.17736, G_{23} = 0.536696, G_{14} = 0.1727503, G_{13} = 0.17736$$
$$G_{24} = 0.1773613, G_{34} = 0.1773613$$

For θ = 30°:

$$G_{12} = 0.1924475, G_{23} = 0.6735778, G_{14} = 0.096224, G_{13} = 0.1924529$$
$$G_{34} = 0.1924576, G_{24} = 0.1924529$$

For θ = 45°:

$$G_{12} = 0.2356953, G_{23} = 0.9714048, G_{14} = -0.028595, G_{13} = 0.2357088$$
$$G_{24} = 0.2357088, G_{34} = 0.2356953$$

For $\theta = 50°$:

$G_{12} = 0.1692328, G_{23} = 0.4266353, G_{14} = 0.2503152, G_{13} = 0.1692420$
$G_{24} = 0.1692420, G_{34} = 0.1692429$

For $\theta = 60°$:

$G_{12} = 0.333037, G_{23} = 1.5232652, G_{14} = -0.1993417, G_{13} = 0.3333476$
$G_{24} = 0.3337476, G_{34} = 0.333037$

For $\theta = 70°$:

$G_{12} = 0.4873140, G_{23} = 2.348359, G_{14} = -0.399141, G_{13} = 0.4872938$
$G_{24} = 0.4872937, G_{34} = 0.487314$

For $\theta = 80°$:

$G_{12} = 0.9597655, G_{23} = 4.75520, G_{14} = -0.916043, G_{13} = 0.9598102$
$G_{24} = 0.9598100, G_{34} = 0.9597658$

It can be observed that as the skew angle is increased, conductor G_{14} decreases and even goes negative for $\theta \geq 45°$. A negative conductance implies heat flow from low temperature to high temperature, violating the second law of thermodynamics. These are special conductors and are required for the correct electrical analog. These conductors are mathematical in nature, and although they violate the laws of nature in a local sense, they preserve the physically realistic solution in a global sense. Negative conductors can be explained physically as being a result of overlapping control volumes, as shown in Fig. 6.13 [24].

One can imagine control volume faces as *AB*, *BC*, *CD* for heat flows between

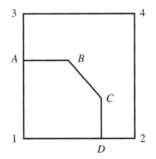

FIGURE 6.13. Control volume faces for conductors 1–3, 1–4, and 1–2.

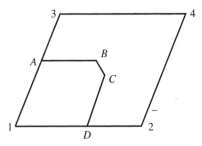

FIGURE 6.14. Control volume faces for a skewed element.

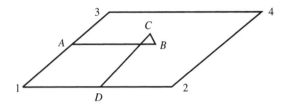

FIGURE 6.15. Control volume faces for a highly skewed element. (Notice the inverted segment *BC*.)

nodes 1–3, 1–4, and 1–2, respectively. As the element is skewed, the length of *AB* and *CD* increases, while length of segment *BC* decreases, as shown in Fig. 6.14. This is why G_{12} and G_{13} increase and G_{14} decreases. As the element is skewed even further, lengths *AB* and *CD* increase further; however, segment *BC* gets inverted (Fig. 6.15). That is why G_{14} becomes negative, while conductors G_{12} and G_{13} continue to increase. These negative conductances are necessary to model the physics of the problem correctly.

ADVECTION AND FLOW NETWORKS

When energy transport occurs because of mass flow stream, it is called advection. For example, when hot fluid is flowing inside a tube, heat (energy) is carried with the fluid into the system. This transfer of energy, positive or negative, by mass flow needs to be accounted for in the thermal model. The resulting heat transfer between nodes 1 and 2 at temperatures T_1 and T_2 is mass flow rate × specific heat of the fluid × difference in temperature

$$Q_{12} = \dot{m}C_p(T_1 - T_2) \tag{6.16}$$

Reverting to the electrical network analogy, the advective conductance G_{12} is equal to mass flow rate multiplied by the specific heat. The specific heat can be evaluated at the average temperature of the two nodes. The conductance that carries the advective energy is a one-way conductance, meaning that the energy can be transported only from the upstream node to the downstream node. All other types of conductances—such as conductive, convective, radiative—are two-way conductances, that is, energy can flow in either direction in those conductors. The matrix with advective conductors therefore is not symmetric. The point iterative SNPSOR algorithm works very well for the system of equations that are diagonally dominant. If there are strong advective flows in the thermal model, the asymmetry of equations may weaken the diagonal dominance and hence slow convergence or result in nonconvergence. One way to ensure convergence in such situations, at the expense of speed, is to underrelax just the advection part of the solution.

In many situations, the mass flow rate is not known, and inlet and outlet pressures or the inlet mass flow and the outlet pressure are known, but the distribution of mass flow rate in the branches of the flow passages have to be determined. For an incompressible, steady state flow condition, pressure drop, elevation drop, and flow velocities are linked by the following equation between points I and J:

$$\Delta Z + \frac{\Delta P}{\rho g} = f \frac{L}{D} \frac{V^2}{2g} \tag{6.17}$$

where $\Delta Z = Z_I - Z_J$, elevation change
$\quad\quad \Delta P = P_I - P_J$, pressure change
$\quad\quad f$ = Friction coefficient
$\quad\quad D$ = Hydraulic diameter of the flow passage
$\quad\quad g$ = Gravitational constant
$\quad\quad A$ = Cross-sectional area of flow passage

Therefore,

$$\rho g \Delta Z + \Delta P = \frac{f L \rho V^2}{D2}$$

Casting the equation in terms of mass flow rate, $\dot{m} = \rho A V$, we get

$$\rho g \Delta Z + \Delta P = \frac{f L}{D} \frac{\dot{m}^2}{2A^2 \rho}$$

$$\dot{m}^2 = \frac{2 D \rho A^2}{f L} \rho g \Delta Z + \frac{2 \rho D A^2}{f L} \Delta P$$

$$\dot{m} = \sqrt{\frac{2 D \rho^2 A^2 g \Delta Z}{f L}} + \sqrt{\frac{2 \rho D A^2}{f L}} \frac{1}{\sqrt{\Delta P}} \Delta P \tag{6.18}$$

Defining the hydraulic conductance or flow conductance as

$$G_n = \sqrt{\frac{2\rho DA^2}{fL}} \ \frac{1}{\sqrt{\Delta P}}$$

$$\dot{m} = G_n\sqrt{\Delta P} \sqrt{\rho g \Delta Z} + G_n \Delta P$$

The pressure drop due to gravity head is

$$\Delta P = \rho g \Delta Z$$

Therefore,

$$\dot{m} = G_n \rho g \Delta Z + G_n \Delta P \tag{6.19}$$

As G_n is a function of ΔP, the equation is nonlinear and has to be solved iteratively. The flow networks generate a nicely banded matrix with a low band bandwidth and therefore can be solved with direct matrix solution techniques quite efficiently. If there are no gravity or other heads (turbine or pump)

$$\dot{m} = [G_n]\Delta P$$

This can be represented in a matrix form as

$$[G_n]\{P\} = \{\dot{m}\} \tag{6.20}$$

The equations are formulated in terms of pressure, and therefore pressure must be known for at least one node point in the system. Once the mass flow rates are determined in the flow network, they are treated as one-way advective conductors in the thermal model.

Example 6.3
Consider a three-pipe system in a series. The inlet pressure is 150,000 Pa, and the outlet pressure is 0 Pa. The diameter and pipe roughness and lengths of the pipe are as shown in Fig. 6.16.
The following FORTRAN program solves the above problem iteratively. The mass flow rate variable is w, pressure is x, and H is the elevation variable. The variable CC stores the $A\sqrt{2\rho/(fL/D)}$ part of the flow network conductance. Recall that the flow conductance G_n is given by

$$G_n = A \underbrace{\sqrt{\frac{2\rho}{fL/D}}}_{CC} \ \frac{1}{\sqrt{\Delta P}}$$

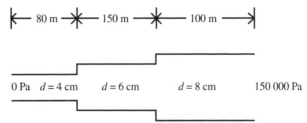

0 Pa $d = 4$ cm $d = 6$ cm $d = 8$ cm   150 000 Pa

Roughness of pipe 1 = 0.2 mm, pipe 2 = 0.12 mm, pipe 3 = 0.24 mm

$Z = 0$ m $Z = 5$ m

FIGURE 6.16. Three-pipe system for Example 6.3.

```
        PROGRAM MAIN
        IMPLICIT REAL*8 (A-H, O-Z)
        DIMENSION A(4,4), B(4), X(4), H(4), XX(4), CC(4), W(4)
C
        H(1) = 5.
        H(2) = 3.4848
        H(3) = 1.2121
        H(4) = 0.0

        DO 550 II = 1,30
        IF(II.EQ.1) THEN

        CC(1) = 0.0374656
        CC(2) = 0.01568377
        CC(3) = 0.0070916

        ELSE

        CC(1) = .0374656/DSQRT(DABS(XX(1) - XX(2)))
        CC(2) = .01568377/DSQRT(DABS(XX(2) - XX(3)))
        CC(3) = .0070916/DSQRT( DABS(XX(3) - XX(4)))

        END IF
C
        A(1,1) = CC(1) + 1.E8
        A(2,1) = -CC(1)
        A(3,1) = 0.
        A(4,1) = 0.
        A(1,2) = -CC(1)
        A(2,2) = CC(1) + CC(2)
        A(3,2) = -CC(2)
        A(4,2) = 0.
        A(1,3) = 0.
```

```
          A(2,3) = -CC(2)
          A(3,3) = CC(2) + CC(3)
          A(4,3) = -CC(3)
          A(4,4) = CC(3) + 1.E8
          A(1,4) = 0.
          A(2,4) = 0.
          A(3,4) = -CC(3)
          AA = 9.81
          B(1) = (150000.*1.E8) + ( CC(1)*1000.*-1.5152*AA)
          B(2) = -(CC(1)*1000.*-1.5152*AA) + (CC(2)*1000.*-2.2727*AA)
          B(3) = -(CC(2)*1000.*-2.2727*AA) + (CC(3)*1000.*-1.2121*AA)
          B(4) = 0. - (CC(3)*1000.*-1.2121*AA)
C
          DO 300 I = 1,4
300       X(I) = 0.
          CALL GAUSSE(A,4)
          DO 100 I = 1,4
          DO 200 J = 1,4
          X(I) = X(I)+ A(I,J)*B(J)

200       CONTINUE
100       CONTINUE

          DO 111 I =1,4
111       XX(I) = X(I) + (1000.*9.81*H(I))
          WRITE(6,*) (X(I), I = 1,4)

          DO 112 I = 1,3
          W(I) = CC(I)*((X(I)-X(I+1))-(1000.*9.81*(H(I+1) - H(I))))
112       CONTINUE

          WRITE(6,*) '----',(W(I), I = 1,3)
550       CONTINUE
          STOP
          END
C
          SUBROUTINE GAUSSE( A, NA)
C
          IMPLICIT REAL*8 (A-H, O-Z)
C----------THIS SUBROUTINE INVERTS SYMMETRIC OR ASYMMETRIC
          MATRIX OF RANK NA
C--       A = MATRIX TO BE INVERTED
C---      NA = SIZE OF MATRIX A
C---      OUTPUT [A] THE INVERTED MATRIX
          DIMENSION A(NA,NA)
          SMALL = 1.E-15
          DO 500 N = 1,NA
          PIV = A(N,N)
          APIV = DABS( PIV)
```

```
            IF ( APIV. GT. SMALL) GO TO 50
            WRITE(6,30) N
30          FORMAT(' ZERO PIVOT( NO ',I2,') FOUND, GAUSSE')
            GO TO 500
50          CONTINUE
            RPIV = 1./PIV
            DO 100 K = 1,NA
            A(N,K) = -A(N,K) * RPIV
100         CONTINUE
            DO 400 I =1,NA
            AIN = A(I,N)
            IF( N. EQ. I) GO TO 350
            DO 300 K = 1,NA
            IF(N . EQ. K) GO TO 300
            A(I,K) = A(I,K) + AIN*A(N,K)
300         CONTINUE
350         CONTINUE
            A(I,N) = AIN*RPIV
400         CONTINUE
            A(N,N) = RPIV
500         CONTINUE
            RETURN
            END
```

Running the program gives the mass flow rate for Example 6.3 as 2.84 kg/sec. The mass flow rates computed from a flow network solution can be used in the thermal analysis. The advective or flow conductance is thermally connected to the solid interface through convective conductance. The advective or flow conductance is a one-way conductance and makes the conductivity matrix asymmetric. The following example illustrates these concepts.

Example 6.4
Air at 25°C is flowing over an edge of a 1 m × 1 m square plate with a mass flow rate of 0.01 kg/sec. The thermal conductivity of the plate is 1 W/m · K. The square plate is modeled as one quad element. There is a heat input of 10 watts at the top left corner (node 3). The convective heat transfer coefficient between the plate edge and the air flow is 10 W/m^2 · K. Assume the specific heat of air to be constant at 1000 J/kg · K. Find the temperature distribution for the system.

The air flow is modeled as an advection bar element between nodes 5 and 6. The advection bar is thermally connected to the nodes on the edge of the plate, i.e., nodes 1 and 2 through the convective boundary condition. There is a heat input of 10 watts at node 3. Nodes 1, 2, 3, and 4 of the quad element are connected by the conductive conductance, which is computed from the finite element matrix formulation for the quad element discussed earlier. Nodes 1 and 5 and nodes 2 and 6 are connected by the convective conductance given by the convective coefficient multiplied by the nodal area. The advective conductance is given by the product of mass flow rate and the specific heat. The advective con-

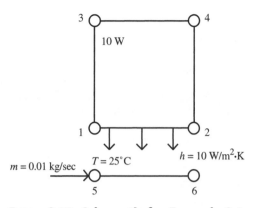

FIGURE 6.17. Schematic for Example 6.4.

ductance is one way, i.e., it only goes downstream from node 5 to 6. Therefore, the conductance network is as follows:

$$G_{1-2} = G_{2-1} = 0.166666$$
$$G_{2-3} = G_{3-2} = 0.333333$$
$$G_{1-3} = G_{3-1} = 0.166666$$
$$G_{1-4} = G_{4-1} = 0.333333$$
$$G_{3-4} = G_{4-3} = 0.166666$$
$$G_{2-4} = G_{4-2} = 0.166666$$
$$G_{1-5} = G_{5-1} = h \cdot A = 0.5(10.0) = 5.0$$
$$G_{2-6} = G_{6-2} = h \cdot A = 0.5(10.0) = 5.0$$
$$G_{5-6} = mC_p = 0.01 \cdot 1000 = 10 \quad \text{one-way conductor}$$

The right-hand side is a null matrix, except for the equation for node 3, for which the heat input is 10.0. Assembling the asymmetric matrix and the right-hand side and solving either by iterative or direct methods, the temperature solution is

$$T_1 = 26.0333 \quad T_2 = 26.45 \quad T_3 = 36.28 \quad T_4 = 36.2 \quad T_5 = 25 \quad T_6 = 25.48$$

PHASE-CHANGE PROBLEMS

This section briefly describes how conduction-based melting and freezing problems are handled in thermal network programs. The numerical methods used for phase-change moving boundary problems can be grouped into two broad classes based on the choice of dependent variable. Methods that use temperature as the dependent variable are called "front tracking schemes." The difficulty with temperature-based formulations is that one must assume a priori the existence and the smoothness of a sharp interfacial surface to even formulate the problem and one has to satisfy the Stefan condition (latent heat energy jump condition) at the interface between the phases. This rules out solving problems with multiple phase changes, those with mushy regions, and problems with phase changes occurring over a temperature range. The second group uses energy as well as temperature as the dependent variable. The method is referred to as "weak formulation" or, more popularly, the "enthalpy method." In this method no a priori assumption is made regarding interface(s) and interface conditions need not be imposed explicitly since they are absorbed as natural interface conditions in the formulation. Solid-liquid regions need not be treated separately; instead, conservation laws are imposed globally, irrespective of the phase. The phases and the interface are determined later from the solution. This leads to a "fixed domain" numerical method without front tracking.

Another method used in thermal network programs is the specific heat method. The capacitors account for latent heat effects at a given temperature as well as normal specific heat effects. To be general, these capacitors should account for multiple phase-change temperatures and latent heat effects within each capacitor. For pure materials, the phase change occurs at a single temperature and not over a temperature range. This results in the internal energy of the material undergoing phase change to cease to be a function of temperature, a circumstance that can cause considerable numerical difficulties. This problem can be circumvented, without much sacrifice in accuracy, by introducing a small temperature band, say 0.1 to 1.0 degrees, over which to spread the effects of latent heat. Introduction of this artificial phase-change band makes internal energy a continuous function of temperature. The introduction of the phase-change band makes the Q versus T curve an S-shaped curve. Newton's method for finding zeros does not work well for S-shaped curves. Therefore, during the iterations, if the temperature of the node trudges into the phase-change band, one has to revert to a bisection algorithm to find the zero of the Q versus T curve. If the phase change occurs over a large temperature range, one can put a bump into the specific heat versus temperature curve in the material properties and the standard conduction-based algorithm can take care of latent heat effects as a part of the normal solution.

RADIATION

Up to this point in our discussions, radiation has been treated as one of the boundary conditions in the thermal analysis. The radiation phenomenon is inherently nonlinear because the thermal energy emitted due to radiation is proportional to the fourth power of absolute temperature. Unlike heat transfer by conduction and convection, radiation requires no medium to transfer energy, and all matter at a temperature greater than absolute zero emits thermal energy. Therefore, heat transfer by radiation is always present and could be the dominant mode of heat transfer at elevated temperatures. Even at moderate temperatures (90–150°C), radiation can contribute significantly to the total heat transfer and should be included in the thermal model.

Radiation heat transfer between a set of surfaces depends on the relative position of these surfaces vis à vis each other, as well as their surface properties (such as emissivity) and their temperature. Surface properties such as emissivity and absorptivity are a function of temperature as well, adding additional nonlinearity into the problem. To compute the radiation heat transfer between these surfaces, one has to introduce the concept of radiation viewfactor (also known as configuration factor or shape factor). A viewfactor, F_{ij}, is defined as the fraction of thermal radiation energy leaving surface i that is incident on surface j. Referring to Fig. 6.18, the viewfactor F_{ij} is defined as

$$F_{ij} = \frac{1}{A_i} \int_{A_i} \int_{A_j} \frac{\cos\theta_i \cos\theta_j}{\pi R^2} dA_i \, dA_j \qquad (6.21)$$

Similarly, F_{ji}, which is the fraction of radiation energy leaving surface j that is incident on surface i, is defined as

$$F_{ji} = \frac{1}{A_j} \int_{A_i} \int_{A_j} \frac{\cos\theta_i \cos\theta_j}{\pi R^2} dA_i \, dA_j \qquad (6.22)$$

From Eqs. (6.21) and (6.22), it follows that

$$A_i F_{ij} = A_j F_{ji} \qquad (6.23)$$

Equation (6.23) is known as the reciprocity relation for viewfactors. This relationship can be used to compute viewfactors from the other known viewfactors and as such cuts down on the number of viewfactors that need to be calculated.

A second important viewfactor relationship is for the surfaces of an enclosure. An enclosure can be defined as a set of surfaces that have a potential to radiate to each other and the open area that the surfaces can potentially see. For an enclosure of n surfaces,

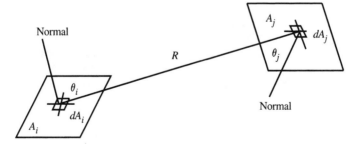

FIGURE 6.18. Sketch defining area elements and angles for a viewfactor.

$$\sum_{j=1}^{n} F_{ij} = 1 \qquad (6.24)$$

This is known as the summation rule for viewfactors.

To calculate radiation heat transfer in an enclosure of n surfaces, a total of n^2 viewfactors need to be determined. Using the summation rule and the reciprocity relationship, only $n(n-1)/2$ viewfactors have to be computed directly [15]. Each finite element in the thermal model that has a radiation boundary condition assigned to it is treated as a surface in the radiation enclosure. As the number of elements (surfaces) in the model increases, the amount of computer resources increases. For geometrically complex models with 1000+ elements (surfaces), the CPU time required to compute viewfactors can be of the same order of magnitude as the CPU required to solve the thermal network.

Analytical expressions for viewfactors are available for very simple geometries [16]. For most situations, the viewfactors have to be computed numerically. Emery et al. [17] have done a comparative study of methods for computing diffuse (surface emissions and reflections are directionally independent) radiation viewfactors for complex geometries. Readers are also referred to books by Siegel and Howell [18] and Modest [19] for various methods available for computing viewfactors. The finite-element-based viewfactor calculation that is used in MSC/Patran/Thermal™ with the hybrid method is briefly addressed later in this chapter.

THERMAL RADIATION NETWORK FOR DIFFUSE, GRAY SURFACES

For the purpose of this book, the discussion is limited to surfaces that emit and reflect radiation independently of direction, i.e., equally in all directions, as well as to surfaces with wavelength-independent surface properties. Such surfaces are referred to as diffuse and gray. The discussion is also limited to surface-to-surface radiation, i.e., the medium enclosed by the surfaces is radiatively nonparticipating. In the previous section, the concept of viewfactor was introduced. Once the viewfactors are calculated, one can proceed to evaluate radiation conductors (or resistors). These conductors are part of the overall thermal network and have to be included in the nodal energy balance.

In an enclosure of n surfaces, the net radiation leaving surface i, q_i, is the difference between the surface radiosity, J_i (total energy leaving a surface by radiation), and surface irradiation, G_i (total radiation incident on a surface), and is given by [15]

$$q_i = \frac{E_{bi} - J_i}{\dfrac{1 - \epsilon_i}{\epsilon_i A_i}} \qquad (6.25)$$

where E_{bi} = Blackbody emissive power of surface $i = \sigma T^4$
$J_i = \epsilon_i E_{bi} + (1 - \epsilon_i)G_i$
G_i = Total irradiation on surface i
ϵ_i = Emissivity of surface i

The driving potential for the heat transfer is $E_{bi} - J_i$. The term $1 - \epsilon_i/\epsilon_i A_i$ is known as the surface or radiosity resistance. The surface conductance is the reciprocal of the surface resistance. In commercial programs such as MSC/Patran/Thermal™, for each surface node that has a radiation boundary condition applied to it and is nonblack ($\epsilon < 1$), a radiosity node is created. These nodes are interior to the program and are created to separate the surface conductances from geometric or viewfactor conductances. The emissivity is evaluated at the average temperature of the surface node and the radiosity node.

The radiation energy leaving surface i is exchanged with each of the other surfaces j in the enclosure based on the viewfactor from surface i to surface j. In other words, all the radiosity nodes exchange radiation with each other through the viewfactor conductance. The heat leaving surface i, q_i, arrives at the radiosity node for surface i and is exchanged with other radiosity nodes. Therefore [15],

$$q_i = \sum_{j=1}^{n} A_i F_{ij}(J_i - J_j)$$

and

$$\frac{E_{bi} - J_i}{\dfrac{1 - \epsilon_i}{\epsilon_i A_i}} = \sum_{j=1}^{n} \frac{J_i - J_j}{(A_i F_{ij})^{-1}} \qquad (6.26)$$

A simple illustration of how Eq. (6.26) translates into a thermal radiation network is given in Fig. 6.19 for an enclosure with four surface nodes.

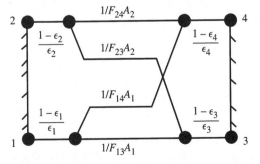

FIGURE 6.19. Thermal radiation network for four surface nodes.

When one is performing energy balances on the surface nodes and radiosity nodes, the heat transfer through radiosity and viewfactor conductances should be computed by multiplying the conductance value by $\sigma(T_i^4 - T_j^4)$.

FINITE-ELEMENT-BASED VIEWFACTORS

As mentioned earlier, in the finite element thermal model, each element that has a radiation boundary condition applied to it is treated as a surface in the radiation enclosure. The hybrid method discussed in this chapter uses the true finite element nodes, and not the element centroids, to construct the thermal network. The element information is used to make the conductors and capacitors. Therefore, the viewfactors also need to be evaluated from node to node, and not from element centroid to element centroid. The finite-element-node-based viewfactor associated with the ith node on surface e to the jth node on surface f is given by the following expression [13]:

$$F_{ij}^{ef} = \frac{\displaystyle\iint_{A^e} \iint_{A^f} \frac{\phi_i^e \phi_i^f (r^{ef} n^e)(r^{ef} n^f) dA^e \, dA^f}{\pi ||r^{ef}||^4}}{\displaystyle\iint_{A^e} \phi_i^e dA^e} \tag{6.27}$$

where e = Surface e ID
$\quad f$ = Surface f ID
$\quad i$ = Node ID on surface e
$\quad j$ = Node ID on surface f
$\quad A^e$ = Surface e area
$\quad A^f$ = Surface f area
$\quad \phi_i^e$ = Finite element interpolation function associated with the ith node on surface e
$\quad \phi_i^f$ = Finite element interpolation function associated with the ith node on surface f
$\quad r^{ef}$ = Vector from a point on surface e to a point on surface f
$\quad n^e$ = Unit normal to surface e
$\quad n^f$ = Unit normal to surface f

The viewfactor given by Eq. (6.27) is computed numerically, using Gaussian quadrature. The vector r^{ef} goes from a point on surface e to a point on surface f. These points are taken to be Gauss points. Thus the viewfactor is calculated from a nodal subarea of element e (area associated with the node) to a nodal subarea of element f. If the vector r^{ef} is obstructed by any other element (surface), then the viewfactor for that nodal subarea is taken to be zero. For each nodal subarea associated with a node, there is a radiosity node, and these radiosity nodes are connected in the thermal network with viewfactor conductors.

SUMMARY

The hybrid method discussed in this chapter brings together the modeling ease and geometric flexibility of the finite element method and the computational efficiency of the finite difference type network solution techniques. It is because of this novel method, which is not covered in any standard numerical heat transfer texts, that the author was motivated to write this primer.

Chapter 7

Software Selection

We have journeyed through the various numerical methods used for conduction-based thermal analysis. There are commercially available design and analysis software packages that incorporate one or more of these numerical methods. These packages range from general purpose (can be applied to solve 80–90% of thermal problems) to very specific (e.g., calculating orbital heating loads, etc.). In the past, access to these packages and the hardware they needed was restricted to the research elite in large corporations, government agencies, and universities. Desktop computing has made major strides in recent years. An engineering workstation or a Pentium™-based personal computer is no longer a luxury; 300–400 megabytes of hard disk space to hold a computer program is no longer considered premium magnetic real estate; and mention of 256 megabytes of memory rarely makes jaws drop in this modern era. A revolution in hardware technology has been coupled with reduced prices in the software industry. As a result, most engineering firms, small or large, can afford to give their designers and engineers access to computer-based simulation tools. At the same time, manufacturers have realized the potential cost savings the engineering analysis tools can bring if they are employed early in the design cycle. Therefore, there is an increasing trend in industry to use the analysis tools throughout the design-to-manufacture product cycle, rather than just as a validation tool or failure analysis tool. Due to this paradigm shift, many design engineers are required to use engineering analysis tools. These design engineers in the past were mainly form, style, or geometry driven. Now they have to be more function driven. Most of the designers have relied on analysis groups to validate their designs in the past. Now they have been asked to validate their own designs, or organizations are forming design teams composed of design engineers and engineering analysts. In the future, these two job functions will merge. As more and more design engineers with little or no experience in engineering analysis start wearing analyst hats, a major question they have to confront is how to go about selecting a computer package to use for their application. The questions one has to ask and resolve are similar regardless of the analysis discipline. As this book is written for thermal engineers, the discussion that follows is written with thermal engineers in mind.

THERMAL ANALYSIS VERSUS CFD ANALYSIS

One of the important questions one has to ask while selecting software to assist in thermal design is the following.

Do I need computational fluid dynamics (CFD) software to analyze the thermal designs or would a conduction-based thermal analysis software without the CFD component suffice?

It is important to understand the differences and similarities between the two disciplines and their domain of applicability. The CFD programs solve the conservation equations of mass and momentum—and, optionally, energy—and can include the solution of flow field, temperature, chemical reactions, concentration, etc. If one is interested in the details of the flow field for design of components, say to predict drag or lift, then one must employ a CFD analysis tool. On the other hand, if one is only interested in the average quantities of the flow, such as average pressure or average mass flow rate at a cross section in a pipe or a channel, then a flow network program is sufficient. When designers have to predict temperatures in their designs, they need a program that solves for the conservation of energy. The flow around the components affects their heat transfer characteristics and therefore temperature; the flow information is required to predict temperatures. As CFD programs predict the flow field, this information can be readily used to solve for the temperature field.

Thermal software, on the other hand, only solves for the conduction-based energy equation (i.e., only in the solid region). The effects of flow field that influence the temperature field are introduced into the thermal model only through the convective boundary condition. The effects of flow on heat transfer, and subsequently temperature distribution, are condensed into a quantity called the heat transfer coefficient. There are geometric configurations and flow conditions for which correlations for the convective heat transfer coefficient exist in the literature. These correlations are obtained either through experiments or with mathematical manipulation of governing equations.

If your design space approximately matches that for which a convective heat transfer correlation exists, then you do not need to solve for the flow field and a CFD solution is not required. Thermal designs can be analyzed with the help of conduction-based thermal software. Conduction-based thermal analysis is a more mature field than CFD analysis, and therefore these programs tend to be easier to use and implement. Most of the commercially available pre- and postprocessors support a thermal analysis program. CFD software, on the other hand, may have its own stand-alone pre- and postprocessing. The finite element meshes created by standard preprocessors may not be compatible with the ones required by CFD software. Also, a higher level of analysis experience and expertise is needed to effectively use CFD as a thermal analysis/design tool. The modeling for CFD is more involved and could be time consuming because one has to model the fluid regions as well as the solid regions. Conduction-based thermal programs do not solve for the flow field because the effects of flow are introduced through the convective boundary condition; therefore, one is not required to model the flow domain. This reduces the model preparation time. It may be better, if one can afford it, to have both CFD and thermal analysis software. CFD can be used to get an idea of the flow field and the convective heat transfer coefficient. The convection coefficient subsequently can be used in a thermal program. Thermal programs are fast to converge to a solution even for highly transient and nonlinear problems involving radiation and can be efficiently used for "what if" thermal design studies.

SELECTING A THERMAL PROGRAM

There are many commercial vendors that offer thermal analysis software. You can select vendors based on industry reputation, size of the company, number of years in business, etc. Before contacting these vendors, you should create a checklist of requirements and questions, such as

- Can the software handle time- and temperature-dependent boundary conditions?
- Can material properties be a function of temperature?
- Are directionally dependent material properties supported?
- Does the software include a built-in radiation viewfactor program?
- Does the software have a library of convection correlations?
- How can one introduce the convection coefficients calculated from a CFD code or experiments into the thermal code?
- How can the temperatures computed from the thermal analysis be mapped for a subsequent thermal-stress analysis?
- Is wavelength-dependent radiation supported?

The above questions are just intended as a guideline; your particular application may have more or less requirements. Besides possessing the ability to analyze your designs from a thermal standpoint, the software should also interface with the design software and other analysis software you use.

Once you have narrowed your search for thermal software, based on the requirements for your application, the next step is to evaluate the reliability of the software. Reliability is the measure of a program's accuracy, i.e., its ability to generate correct answers when given the correct input. Most software vendors have a suite of test problems that they use to determine the accuracy of their software. These test problems are usually a mix of textbook problems, as well as complex, real-world problems. You may want to ask the vendor to demonstrate how to model and solve problems typical to your application. It is advisable not to put the software into a production environment until it has passed your acceptance test.

Besides being accurate, the software must be easy to use. All vendors say that their software is intuitive and easy to use. This is a subjective judgment and unless you test drive the code, you can not assess its usability/ease of use. To effectively use the software in production mode requires training and technical support. Therefore, it is important to evaluate the quality of the technical support, its accessibility, and its frequency and quality of training. Lastly, the software should be portable between various operating systems and should be supported on your current and future hardware platforms.

Thermal analysis software allows an engineer to simulate the real-world behavior of designs in the computer. The results can be used to verify, as well as optimize, designs before prototypes are built. However, this does not mean that one can completely eliminate costly physical testing. The analysis tools can be effectively used to minimize the number of physical tests and prototypes. Computer simulation gives an idea of how a system or a design will perform in a given environment; testing, on the other hand, shows how it performed. One should be aware that just as there are uncertainties in physical testing, there are uncertainties in computer analysis. The computed solution is only as accurate as the accuracy of the input. Inaccuracies in describing boundary conditions or material properties will lead to a numerically accurate wrong answer. So how can a person know whether he or she has a good computer solution? Here are a few tips:

- Check the accuracy of the boundary conditions and material properties and their unit consistency.
- Check whether the total heat input reflected in the solution (output) is accurate.
- Some programs report an energy balance in the solution. For a steady state thermal solution, the energy balance should be a small percentage of the total heat input (ideally close to 0.0).
- First run an analysis for a relatively small model (# of nodes) to check the accuracy of the input data. Use the results of this simulation to identify areas of high temperatures and temperature gradients. Refine the mesh in these areas and rerun the analysis. Stop the refinement when the percent change in termperatures is small.

Finally, there is no technology available that can take the place of engineering judgment.

References

1   Ozisik, M. N., 1980, *Heat Conduction*, New York, John Wiley & Sons.

2   Rosenow, W. M., Hartnett, J. P., and Ganic, E. N., Editors, 1985, *Handbook of Heat Transfer Fundamentals*, New York, McGraw-Hill Book Company.

3   Press, W. H., Flannery, B. P., and Teukolsky, S. A., 1986, *Numerical Recipes*, Cambridge University Press.

4   Patankar, S. V., 1980, *Numerical Heat Transfer and Fluid Flow*, New York, McGraw-Hill Book Company.

5   Zienkiewitz, O. C., 1977, *The Finite Element Method*, 3rd Edition, New York, McGraw-Hill Book Company.

6   Reddy, J. N., 1984, *An Introduction to Finite Element Method*, New York, McGraw-Hill Book Company.

7   Baker, A. J., 1983, *Finite Element Computational Fluid Mechanics*, New York, Hemisphere Publishing.

8   Pepper, D., and Heinrich, J. C., 1989, *Finite Element Heat Transfer*, American Institute of Aeronautics and Astronautics (AIAA) Home Study Course.

9   Chainyk, M., 1994, *MSC/NASTRAN Thermal Analysis Handbook*, Pasadena, CA, MacNeal-Schwendler Corporation.

10   Thermal Engineering with Patran/Thermal™, 1994, Course notes for Pat 312, Pasadena, CA, MacNeal-Schwendler Corporation.

11   Ketkar, S. P., 1993, "Iterative and Direct Solution Methods in Thermal Network Solvers," *International Communications in Heat and Mass Transfer*, Vol. 20, No. 4, pp. 527–534.

12   Hughes, T. J. R., 1977, *Computer Methods in Mechanics and Engineering*, Vol. 10, pp. 135, North Holland Publishing Company.

13   MacNeal-Schwendler Corporation, 1992, *MSC/Patran/Thermal™ Users Manual*, Pasadena, CA, MacNeal-Schwendler Corporation.

14   Hageman, L. A., and Young, D. M., 1981, *Applied Iterative Methods*, New York, Academic Press.

15   Incropera, F. P., and DeWitt, D. P., 1985, *Introduction to Heat Transfer*, New York, John Wiley & Sons.

16   Howell, J. R., 1982, *A Catalog of Radiation Configuration Factors*, New York, McGraw-Hill Book Company.

17   Emery, A. F., Johansson, O., Lobo, M., and Abrous, A., 1991, "A Compara-

tive Study of Methods for Computing the Diffuse Radiation Viewfactors for Complex Structures," *ASME Journal of Heat Transfer*, Vol. 113, pp. 413–422.

18   Siegel, R., and Howell, J. R., 1981, *Thermal Radiation Heat Transfer*, New York, Hemisphere Publishing Corporation.

19   Modest, M. F., 1993, *Radiation Heat Transfer*, New York, McGraw-Hill Book Company.

20   Smith, G. D., 1985, *Numerical Solution of Partial Differential Equation: Finite Difference Methods*, London, Clarendon Press.

21   Carlslaw, H. S., and Jaeger, J. C., 1959, *Conduction of Heat in Solids*, London, Oxford University Press.

22   Crank, J., 1975, *The Mathematics of Diffusion*, Clarendon Press.

23   Zarda, P. R., Anderson, T., and Baum, F. (Martin Marietta) London 1988, *FEM/SINDA, Combining Strengths of MSC/NASTRAN,TM SINA,TM Supertab,TM and Patran,TM for Thermal and Structural Analysis*, Pasadena, CA, MacNeal-Schwendler Corporation.

24   Manteufel, R. D., 1987, "Analagous Finite Element/Difference Method for the Heat Diffusion Problem," M.S. Thesis, University of Texas at Austin.

Index

A

Advection, 81
 flow networks and, 81–87

B

Backward difference expression, 10
BEM (boundary element method), 2
Boundary conditions, 6–8
 natural, 45
Boundary element method (BEM), 2

C

CAE (computer–aided engineering), 1
Capacitor, defined, 74
CARE (connectivity and resistor equivalent) technique, 2
CFD, *see* Computational fluid dynamics analysis
Choleski method, 71
Computational fluid dynamics (CFD) analysis, 96
 thermal analysis versus, 95–96
Computer-aided engineering (CAE), 1
Computer-aided numerical thermal analysis, vii
Conductances, negative, skewed meshes and, 78–81
Configuration factor, *see* Viewfactors
Connectivity and resistor equivalent (CARE) technique, 2
Conservation principle, 21
Control volume method, 1, 21–31
 boundary conditions, 23–25
Crank-Nicholson method, 30, 42

D

Dirichlet boundary condition, 7, 23–24

E

Element centroidal approach, 60–61
Enthalpy method, 88

F

FDM, *see* Finite difference method
FEM, *see* Finite element method
Finite difference method (FDM), 1, 9–19, 33
 direct solution method, 13–14
 explicit method, 18
 implicit method, 19
 iterative solution technique, 14–17
 sample FORTRAN program for, 15–17
Finite-element-based viewfactors, 92
Finite element method (FEM), 2, 33–57
 boundary conditions, 35–36
 hexahedron element in, 54–55
 quadrilateral element in, 48–52
 steps in, 33
 tetrahedron element in, 52–54, 56–57
 three-dimensional elements in, 52–57
 triangular element in, 43–48
 two-dimensional elements in, 42–52
Finite element to resistor-capacitor method, 63–70
Finite volume method, *see* Control volume method